Engineeri DESIGN
A REVIEW

MW01614402

Edited by

SVEN G. BILÉN

School of Engineering
The Pennsylvania State University

 McGraw-Hill Primis
Custom Publishing

Boston   Burr Ridge, IL   Dubuque, IA   Madison, WI   New York   San Francisco   St. Louis
Bangkok   Bogotá   Caracas   Lisbon   London   Madrid
Mexico City   Milan   New Delhi   Seoul   Singapore   Sydney   Taipei   Toronto

McGraw-Hill Higher Education
A Division of The **McGraw-Hill** Companies

**ENGINEERING DESIGN
A REVIEW**

McGraw-Hill's Primis Custom Publishing consists of products that are produced from camera-ready copy. Peer review, class testing, and accuracy are primarily the responsibility of the author(s).

890 QSR QSR 098765

ISBN 0-07-281900-6

Sponsoring Editor: Margaret Hollander
Production Editor: Kathy Phelan
Printer/Binder: Quebecor World

Brief Contents

Contents

The materials in this book were initially collected to support an introductory engineering design course. Hence, many students may have already used the concepts in this book, most likely through an introductory course on engineering design. However, I realize that not all of you have seen this material in an organized format. Therefore, I think it appropriate to provide it again as a refresher for some and as a resource for all as you approach your capstone design project.

If you took it, the introductory course in engineering design allowed you to experience the design process early in your studies. This was important because it gave you the framework to understand how all the various pieces of engineering knowledge and skills you have acquired fit together. Now, as you are finishing your undergraduate career, the senior "capstone" design class provides an excellent method for pulling all the pieces together. However, it is also important to remember as your pursue your design, that the design process is still applicable—indeed necessary. For example, take care not to pursue blindly the first idea that comes to mind. Rather, determine first what the customer needs are, then develop an array of alternative designs, and employ a selection process to come up with the most "optimal" solutions. Once you have developed your solution through an iterative process, you must test it and finally, communicate it. Reflecting on the process throughout will help you hone your skills so that your next design effort runs more smoothly.

I wish you the best in your capstone design experience and in your future engineering career!

Sven G. Bilén
Assistant Professor of Engineering
Design and Electrical Engineering

Chapter

An Introduction to Engineering Design

1

What is meant by *design* or, more specifically, by *engineering design*? When asked to name a designer, most people will suggest some painter, sculptor, or fashion designer. These are indeed practitioners of design, but in a different sense than what is meant by engineering design. The difference is that "artistic" designers typically execute their designs with few customer interviews, no analysis, often no prototypes, and no detailed drawing set. Rather, they have an intuitive understanding of what will work or what looks good. In a sense, the design of the artifact is not separate from the process of making it. Hence, this type of design may be referred to as *craftsmanship*.

Engineering design, on the other hand, is a process that requires the use of engineering elements and tools to complete the design task. In engineering, a systematic effort is employed to go from problem statement to solution and ultimately to implementation. Design—the process of providing solutions to societal needs through the conversion of resources—is the essence of engineering. Although the craftsmanship approach occasionally may be attempted in engineering, engineering design techniques are needed to consistently and quickly bring reliable, quality, and desirable products to market.

The purpose of this book and corresponding course is to introduce the first-year engineering student to this "essence of engineering": design. In the past, learning the concepts and principles of design was often reserved for a senior "capstone" design class. By waiting until senior year it was felt that students had developed enough engineering analysis skills and the design course acted as a way of pulling all the pieces together. This rationale is an important one and thus you most likely will take such a capstone design course as a senior. However, because design is so integral to what we do as engineers, it

"Scientists investigate that which already is; engineers create that which has never been."

Theodore von Kármán, aerospace pioneer

"Engineering is the art of modeling materials we do not wholly understand, into shapes we cannot precisely analyze so as to withstand forces we cannot properly assess, in such a way that the public has no reason to suspect the extent of our ignorance."

A. R. Dykes, British Institution of Engineers, 1976

"Engineering is sometimes thought of as applied science, but engineering is far more. The essence of engineering is design and making things happen for the benefit of humanity."

Martha Sloan, Chair, American Association of Engineering Societies

"Engineering design is the process of devising a system, component, or process to meet desired needs. It is a decision-making process (often iterative), in which the basic sciences and mathematics and engineering sciences are applied to convert resources optimally to meet a stated objective. Among the fundamental elements of the design process are the establishment of objectives and criteria, synthesis, analysis, construction, testing, and evaluation."

ABET 2000

is important that you are exposed to it early in your studies. Indeed, it is hoped that by experiencing the design process early on, you will be able to understand how all the various pieces of engineering knowledge and skills you will be acquiring fit together. When the aspects of design evidence themselves in your courses, you will have an appreciation for them. Although it is important to learn the process of design, it is also important that you learn design by doing design. This course will offer you that opportunity. It must be mentioned, however, that this course is meant to be an introduction to engineering design, a first exposure, and not a complete treatment.

To this point in your academic career, you have been trained to seek the "right answer." You will see in this course that there will not be just one right answer to a design problem, but an array of acceptable or optimal answers. How, then, do engineers find these answers? By employing the engineering design process.

1.1 | THE ENGINEERING DESIGN PROCESS

As referred to above, one of the aspects that differentiates engineering design from craftsmanship is the employment of a process. As an engineer, you will have to understand this engineering design process and be adept in the use of various engineering design tools. This is true even if you find yourself as a sales or manufacturing engineer. No longer is design relegated to research and development (R&D) department, it exists throughout the engineering organization.

There are many models of the design process, but all have certain features in common. All design efforts involve systematic problem solving, they are cyclical and iterative, and they have a start and a finish. The engineering process may be thought of as having roughly four phases (Figure 1.1):

- Defining the Problem
- Developing Concepts or Solutions
- Evaluating Solutions
- Communicating and Implementing the Design

It is also important to note that because the design process is iterative, often the designer must return to a previous phase. We will now examine each of the design phases in more detail below.

Defining the Problem

The design process starts with defining the problem. In some cases, the design problem may be well defined, perhaps as part of an obvious need suggested by customers. In many cases, however, the designer is faced with a poorly defined problem, yet the solution he or she develops obviously must be well defined. The first step, then, is for the designer to understand the problem. Only then can he or she begin finding the solution. In some cases, the problem and solution concepts are developed side-by-side so that the solution concepts help the designer understand the real problem. We are often so ingrained in one manner of thinking that we close ourselves off from finding a better solution. Indeed, there are two opposite tendencies that must be reconciled. On the one hand, you want to define the problem as definitively as is possible; on the other hand, your problem definition must remain open enough that you do not preclude from consideration all viable solutions.

Defining the problem also may be thought of as "understanding the opportunity." In general, the success of a for-profit company depends on the creation of products that meet the needs of their customers in a timely, cost-effective manner. To know what their customers want, companies often engage in what is known as a *customer needs assessment*. Yet, even if the company develops an

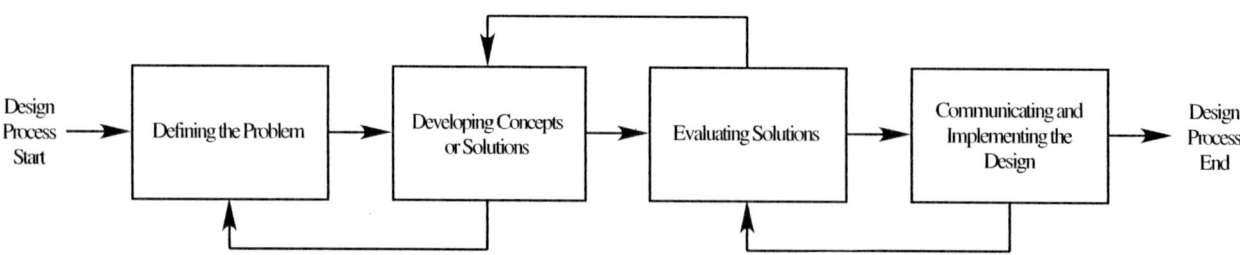

Figure 1.1
A simple four-stage model of the engineering design process

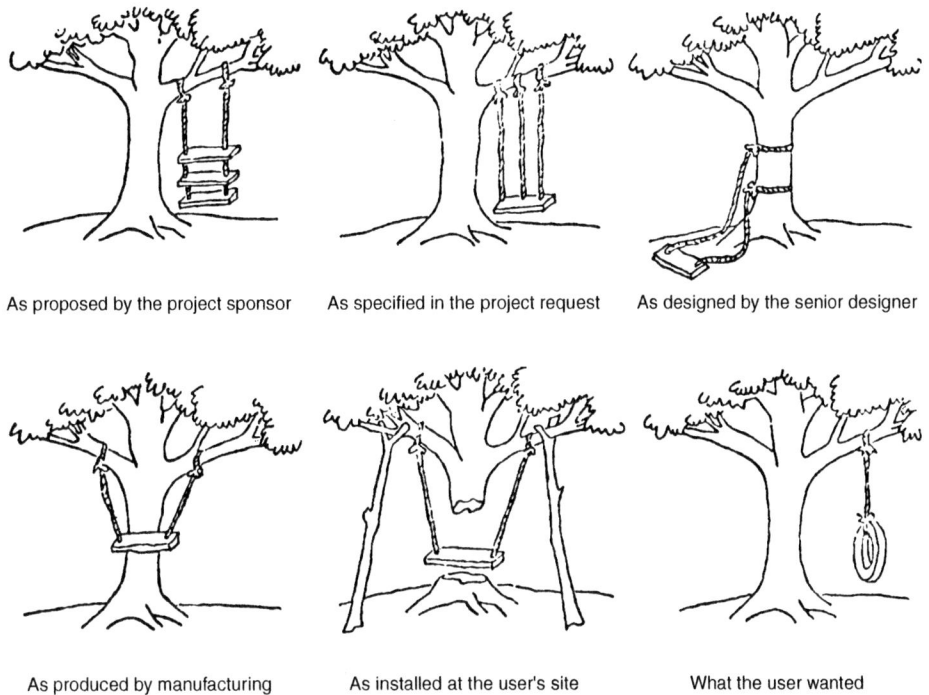

As proposed by the project sponsor   As specified in the project request   As designed by the senior designer

As produced by manufacturing   As installed at the user's site   What the user wanted

Figure 1.2

The solution to the problem often depends on who writes the problem definition. Reprinted from *Engineering Design: A Materials and Processing Approach*, 3/e, by George E. Dieter (2000, The McGraw-Hill Companies, Inc.).

excellent product, if they are late to market or the product is released in the wrong color, they may have missed the opportunity the market provided them. Some questions you might consider when defining a problem include: What solutions already exist? What requirements must the solution meet? What constraints must the solution adhere to (time, cost, weight, size, environmental, safety, social, political, etc.)?

The problem statement must be understandable to all members of your design team. It is not likely, and probably undesirable, that every team member be an engineer, so your problem definition must take this into consideration. That is, it should be presented as clearly and concisely as possible and take into account that other members of your team will have different backgrounds. It is often a good idea once the definition is prepared to have the design's stakeholders (engineering, marketing, customers, etc.) interpret it. Even if you think the problem definition is clear, Figure 1.2 shows how the problem definition may appear to the other stakeholders!

Developing Concepts or Solutions

Once the design problem is understood, the engineer begins by generating a set of conceptual solutions. These concepts are then evaluated with respect to the design requirements and refined. As mentioned above, typically there will not be one "right answer," rather a spectrum of possible alternatives. Hence, it is important that as many concepts as possible are generated, such that an optimal[1] solution can be found by assessing the pros and cons of each. It may be that combining features from several different concepts generates a more optimal solution. As an engineer, you will find yourself having to make trade-offs between the various design parameters and requirements. That is, you may find that an ideal solution to one part of your design means a less-than-optimal—or even an impossible—solution in another part.

[1] Note the use of the word "optimal" and not "optimum," which would suggest convergence to a "right answer."

Figure 1.3
Dilbert by Scott Adams. Reprinted by permission of United Features Syndicate, Inc.

In this stage you will concern yourself with generating as many solutions as possible. You might begin with the process of "brain storming", or a more methodical approach like *concept generation,* which determines design variations by decomposing the problem into subsystems and combining all possible variants of each subsystem. It is important that you not discard potentially good ideas by putting artificial restrictions on your output.

Evaluating Solutions

Evaluating design solutions is an essential part of the design process. Before the item is manufactured it is important that it be checked thoroughly: ensuring that components fit together properly, ensuring the design will withstand design loads, evaluating the difficulty of assembly, etc. Usually, the evaluation stage will find problems in the proposed design and refinements need to be made. The refinement process is often the most time-consuming portion of the design process. Adjustments to one part of the design generally require repeating of set of analyses for the entire design and may lead to unforeseen problems. The process of refinement and repeated analysis is called *iteration*, a common feature of designing.

It is important to select the right kind of evaluation mechanism, whether it is an analytical model, a computer analysis, a prototype for testing, or some other modeling process. When generating the model, it is critical that a given design be evaluated against the original criteria established at the beginning of the design process (and possibly subsequently refined). These criteria may have been established by the client or customer, or may be governed by industry standards or government regulations. The designer may also have other requirements that must be met,

sometimes called "design-for-X" methods, such as design for the environment or design for manufacturing.

Iteration: The Social Dynamics of Design

In Figure 1.1, we have shown how design is a process with stages and have indicated some backward loops of iteration between stages. Design is continually affected by new inputs of information, people, ideas, and changed objectives from the client. For routine design efforts, such as for a car door on a standard model, these inputs may not occur frequently. However, when designing an innovative new product, nothing stands still for long. The iteration required to absorb the new information or objectives may be minor or require a completely new approach. In addition, iterating the design will generally continue even after the product is on the market, since redesign to remain competitive, based in part on feedback from users, is a major activity for design engineers. Good design engineers cannot afford to be dogmatic or defensive about their designs. They must continually reevaluate the tradeoffs of what they are doing in the light of the new data and ideas. That is why documenting the decision making process with such tools as selection matrices is very important. The documentation allows you to see the implications of any changes very clearly. One of the tradeoffs, of course, is time to market, and, at some point, the pressure to stop making changes and move forward will become very strong. But once the process is grasped, design is a very exciting and rewarding activity.

Communicating and Implementing the Design

At the "end" of the design process, the engineer must be able to communicate the design. In this sense it might be said that the design is the description, the drawings, the report, etc. Thus, one of the most important tasks of the engineer is the production of the final design description in a form that is understandable to those who will be approving and implementing the design. This phase of the design process pulls together all the pieces such that your design is presented in a complete and compelling manner. Your presentation of the design must convince others (e.g., management, your client, the Board of Directors your professor and classmates in an academic setting) that the design is feasible and meets the requirements established at the beginning of the design process. Only then can your design be implemented.

Perhaps the most widely used form of communication is the engineering drawing. You will need to learn the "language" of engineering graphics communication in order to unambiguously present your final design solutions. In addition, the techniques you learn can be applied throughout the design process, from first sketches to detailed assembly drawings. Final drawings are normally done with a computer aided design (CAD) package. As such, one might argue that learning manual techniques are no longer important. On the contrary, they are still necessary, particularly at the early stages of the design process where is it important to get your ideas quickly onto paper. In addition, learning manual drawing techniques allows you to determine whether or not what the CAD software produces is correct. As yet, there is no CAD package that produces flawless multiview drawings, fully dimensioned and annotated.

1.2 | ETHICS, TECHNOLOGY, AND DESIGN[2]

While engineers necessarily devote much time to the difficult technical tasks involved with designing new technology, they must be aware that technology itself is part of a much broader social process. Technology is human behavior that transforms society and transforms the environment. Design is the cornerstone of technology. Design is how we solve our problems, fulfill our needs, express our values, shape our world, change the future, and create new problems and new opportunities for present and subsequent generations of all species. Design is quintessentially

an ethical process. As mentioned above, we always "design for," such as designing for society, the environment, safety, assembly, disassembly, manufacturability, profit, jobs, consumer satisfaction, national security, and so on. From extraction to disposal in the life cycle of a product, the design process is where the most important decisions are made or reflected. These decisions determine most of the final product cost, and also most of the ethical costs. They also determine the product's benefits and to whom they accrue. Hence, design is already essentially an ethical process, and the curriculum is not intended to add ethics to that which is only technical, but rather to expand the ethical imagination by revealing more clearly the ethical aspects of design and exploring various alternatives and viewpoints about the design options and the problem(s) the design seeks to solve. Whether viewed from an ethical or technical perspective, creativity that generates many options from which to choose is very advantageous for good design.

Most approaches to ethics focus on individual behavior, often referring to the codes of ethics of engineering societies, case studies of major engineering failures, and conflicts of interest in the professional practice. However, almost all technology is produced through a series of decisions that are made collectively and the social arrangements for making decisions can vary and, as they do, the technical and ethical consequences will vary as well. For example, in recent years industry has moved more toward concurrent product design and development teams—which include many specialists such as design engineers, manufacturing engineers, and sales staff—that stay with the product through the design and manufacturing stages and perhaps through the use stage since much design is redesign to improve existing products and manufacturing processes. Concurrent engineering replaces the traditional sequential process of using separate teams at each stage and it leads to better quality products and more efficient manufacturing processes, both of which obviously have ethical connotations. Similarly, we have become much more user centered in our designs and usually do customer needs assessment prior to or as part of design. This is a very creative change in the social arrangements for making design decisions that makes a product much more likely to do well in the market because it is much more likely to meet social needs. So, engineers need to look closely at the social processes they are using for teamwork, project management, client relations, customer needs assessment, and so on.

[2] Devon, Richard F., "Towards a Social Ethics of Technology: The Norms of Engagement," *Journal of Engineering Education,* Vol. 88, No. 1, pp. 87–92, January 1999.

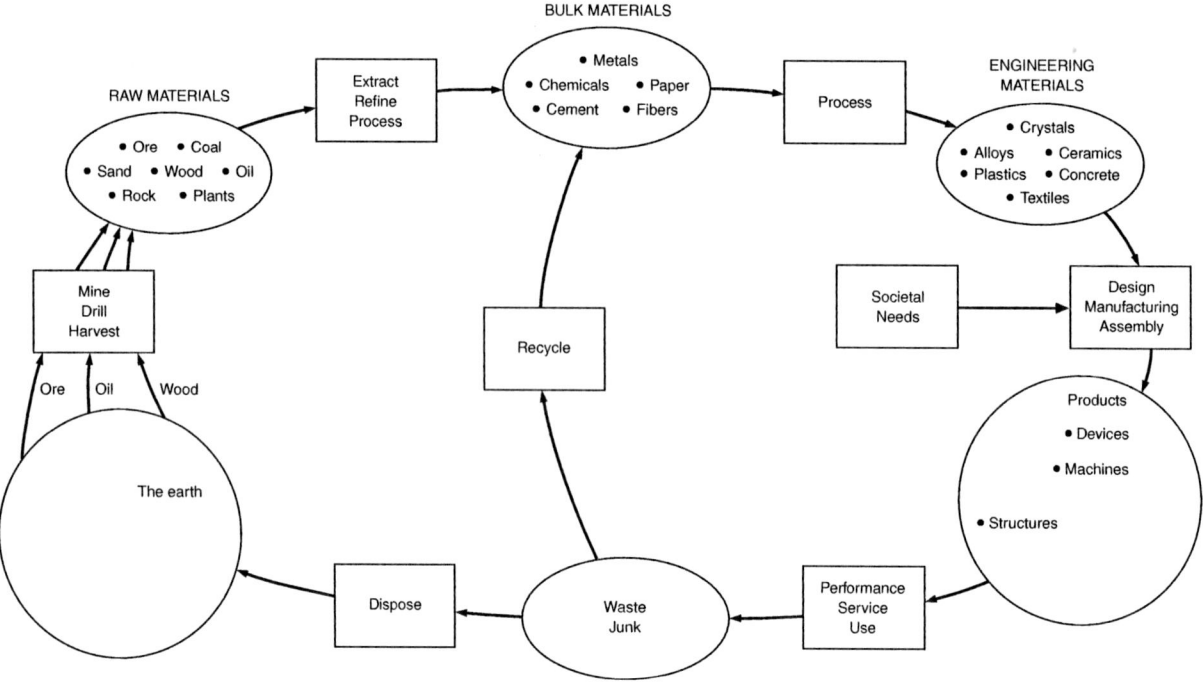

Figure 1.4

The materials or product life cycle, showing that societal needs drive the process. Reproduced from "Materials and Man's Needs," National Academy of Sciences, Washington, DC, 1974.

In recent decades we have become more aware of the fact that we live in a closed ecosystem (plus solar energy coming in and heat loss going out) that has limited resources in energy and materials and a limited capacity of the environment to absorb the waste products that we generate through technology (Figure 1.4). As engineers, we must consider the entire product life cycle (life cycle assessment) and think of ways to make technology, and hence society, environmentally sustainable. It is projected, for example, that oil will run out around mid century and a diverse set of alternatives must be developed, such as solar, hydrogen, and fuel cells.

Engineers must recognize the transformative and collective nature of technology and focus as much on what makes a technology "good" as we do on what makes an engineer "good." This includes far more than the engineer in the product design and development process. All those who help to shape the decision-making process are part of the ethics of design. In addition, ethical reflections must be applied to the entire product life cycle. The curriculum should help students look upstream and downstream in the transformations caused by the product's life cycle and it should help raise the cognizance of decision-makers

with respect to the stakeholders involved and examine who is driving the production of this product or service and why.

In addition to the curricula materials on ethics in this book, students are encouraged to value their general education courses that add human and social perspectives to their technical knowledge. There should be no barriers between these two modes of thinking if design is to be done well.

1.3 | ENGINEERING DESIGN TEAMS

The idea of the lone inventor resides deep in the American psyche: witness Henry Ford, Thomas Edison, or Eli Whitney. Although this perception is perhaps formed by the public's search for "heroes," the reality is that few products today are developed by a single person—if they ever were. As an engineer, you will be part of a project team, a collection of people with various backgrounds that come together to meet the goal of developing a successful product. Some members may be involved with the project from start to finish; others may join during specific phases of the project. In addition, there is often what is known as a *core team* and an *extended team*. The core team is frequently a small group that works on the project daily—although they may have other on-going projects—and is very familiar with its various aspects. The extended team may include employees in finance or sales, as well as those at partner companies, suppliers, and consulting firms.

In this course you will be part of a design team. You and your teammates will compose the core team working on several design projects. It is important that your team move through the inevitable conflicts, learn to communicate, determine how to make decisions, and organize your efforts. When you have done this, you will find that your team is effective and functions as more than simply a sum of the individual members.

Because design is a process with a certain set of steps, it is a something that can be managed by your team. Learning how to manage a complicated design project is an important part of your training as an engineer. You will learn that how well a project is managed determines how successful it will be. There are techniques for project management—the DSM, the Gantt chart, the PERT chart, and the critical path method—that are based on understanding and representing each of the tasks involved and determining their dependencies to each other. A good project plan will allow your team to manage the required tasks given the limited set of resources available to you such as time and personnel.

1.4 | ENGINEERING DESIGN TOOLS

As mentioned, perhaps the most widely used method for communicating your designs is the engineering drawing. This course will emphasize graphics communication, the "language" of the engineer, and one the most important tools available to the engineering designer. You will use engineering graphics to communicate technical ideas and your design solutions. Because it is a language, there are definite rules that must be mastered and followed in order to communicate effectively with other engineers, machinists, assembly line workers, etc. Once you have mastered the language, you will find that it helps you to visualize your designs and you will begin to use graphics in all stages of the design process, from concept to detailed drawing (Figure 1.5).

You will be filling your "engineering toolbox" with general and discipline-specific techniques that you will learn in your many engineering classes. These may include topics such as circuit analysis, structural analysis, thermodynamics, electromagnetics, reaction, nuclear theory, etc. In this course, we will focus on techniques and tools that are less discipline specific but, rather, that can be applied to the spectrum of engineering disciplines. You will be exposed to the spreadsheet as a tool for analysis and design. This software tool has quickly become ubiquitous in design because of its versatility and its near universal acceptance and availability.

1.5 | ROAD MAP TO THIS BOOK

This book has been developed to provide the beginning engineering student with the process and the tools needed for successful engineering design efforts. Because this course and book provide you with an introduction to engineering design, it is our hope that you will use this as a framework for your future studies.

Part 1 of this book presents you with an overview of the design process. The chapters of this part introduce development processes and organizations (Chapter 2), how customer needs are identified (Chapter 3), how product specifications are established (Chapter 4), and the techniques of concept generation and selection (Chapters 5 and 6).

Part 2 of this book examines the role of ethics in engineering design. We will examine the idea of engineering as social experimentation (Chapter 7), look at what are your workplace responsibilities and rights (Chapter 8), and what are some of the global issues concerning engineering (Chapter 9).

Part 3 of this book examines the nature and management of project teams. We will look at the concept of teamwork (Chapter 10) and some methods for developing teamwork skills (Chapter 11). Fundamental to successful design projects is effective management of them; hence, project management techniques are presented (Chapter 12).

Part 4 of this book presents the design tools of engineering graphics. As it is important to communicate your designs effectively, this part presents the techniques for graphics communication (Chapters 13 through 20).

Part 5 of this book presents the spreadsheet as a design tool. The spreadsheet has quickly established itself as one of the more effective tools of analysis and modeling. The fundamentals are presented first (Chapter 21), and then methods for graphing (Chapter 22) and fitting (Chapter 23) are presented. Many additional techniques may be employed with spreadsheets, but these chapters present two of the most universal needs.

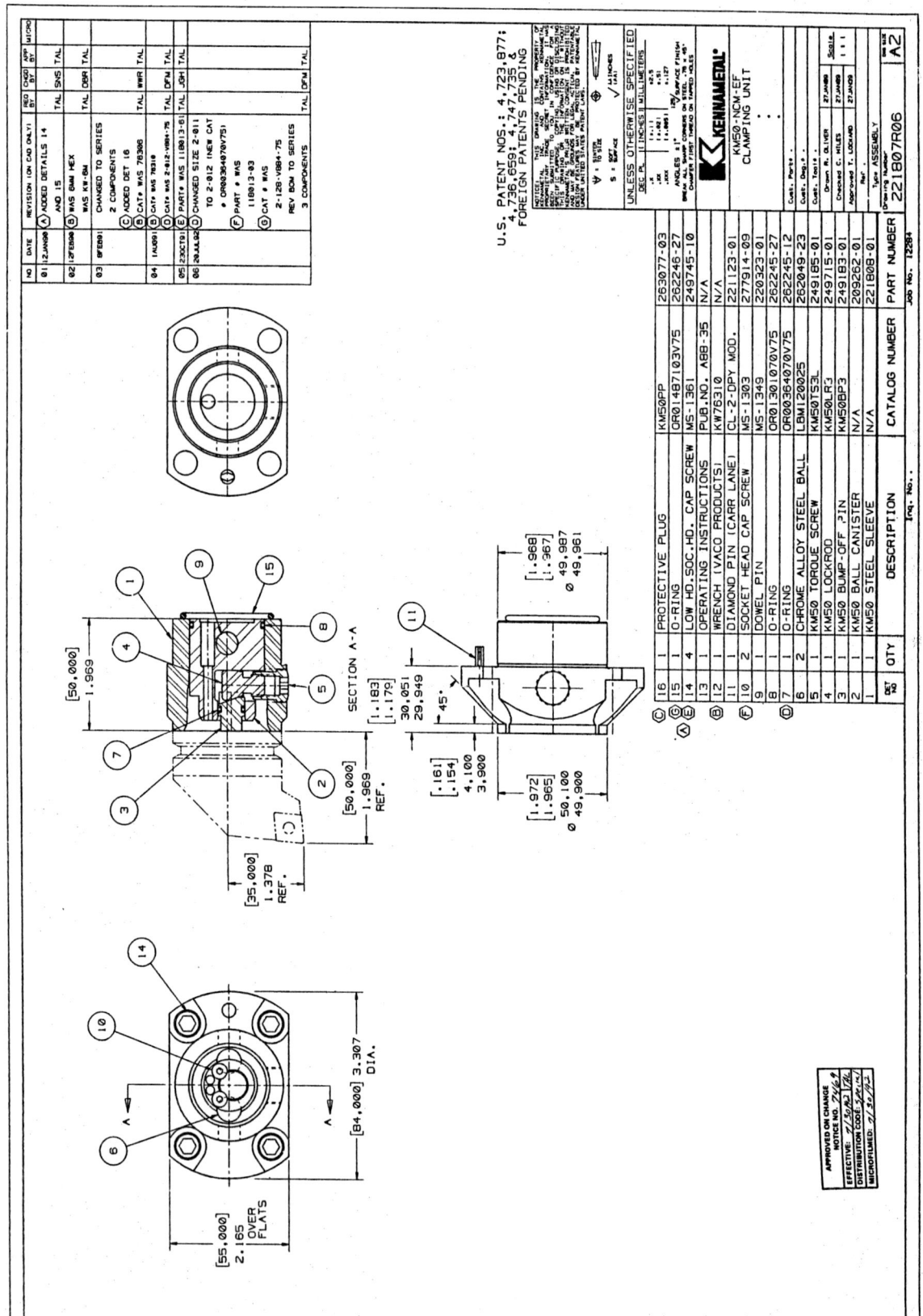

Figure 1.5

An example of a working assembly drawing   (Courtesy of Kennametal.)

References and Bibliography

Cross, Nigel, *Engineering Design Methods: Strategies for Product Design*, Third Edition, John Wiley & Sons, Chichester, 2000. This book summarizes many of the different models that have been published on the design process.

Dieter, George E., *Engineering Design: A Materials and Processing Approach*, Third Edition, McGraw–Hill, New York, 2000. A book on engineering design that places emphasis on areas that the author felt were not emphasized in other texts, those being materials and how they are processed to achieve a design goal.

Horenstein, Mark N., *Engineering Design: A Day in the Life of Four Engineers*, Prentice–Hall, Upper Saddle River, NJ, 1998. This book explores the engineering design process through the activities of four fictional engineers on a design team.

Otto, Kevin, and Kristin Wood, *Product Design: Techniques in Reverse Engineering and New Product Development*, Prentice–Hall, Upper Saddle River, NJ, 2001. This book looks at the product development process through the technique of reverse engineering. By examining good designs, principles of the design process are extracted and emulated.

Ryan, John C., and Alan Thein Durning, *Stuff: The Secret Lives of Everyday Things*, Northwest Environment Watch, Seattle, WA, 1997. This book looks at consumer products throughout the manufacturing process, showing how, for example, your cup of coffee may impact the pristine Siberian wilderness!

Ulrich, Karl T., and Steven D. Eppinger, *Product Design and Development*, Second Edition, McGraw–Hill, New York, 2000. Although several chapters of this text are included in this book, the authors present additional useful resources such as design for manufacturing and product architecture.

Part

I

The Design Process

Chapter

Development Processes and Organizations 2

The Capital Equipment Division of AMF Bowling is the leading manufacturer of bowling equipment, including pin spotters, ball returns, and scoring equipment. An AMF ball return product is shown in Figure 2.1. The general manager of the division asked the engineering manager to establish a well-defined product development process and to propose a product development organization that would allow AMF to compete effectively over the next decade. Some of the questions AMF faced were:

- Is there a standard development process that will work for every company?
- What role do experts from different functional areas play in the development process?
- What milestones can be used to divide the overall development process into phases?
- Should the development organization be divided into groups corresponding to projects or to development functions?

Figure 2.1
A ball return, one of AMF Bowling's products.

(Courtesy of AMF Bowling.)

This chapter helps to answer these and related questions by presenting a generic development process and showing how this process can be adapted to meet the needs of particular industrial situations. We highlight the activities and contributions of different functions of the company during each phase of the development process. The chapter also explains what constitutes a product development organization and discusses why different types of organizations are appropriate for different settings.

2.1 | A GENERIC DEVELOPMENT PROCESS

A process is a sequence of steps that transforms a set of inputs into a set of outputs. Most people are familiar with the idea of physical processes, such as those used to bake a cake or to assemble an automobile. A *product development process* is the sequence of steps or activities which an enterprise employs to conceive, design, and commercialize a product. Many of these steps and activities are intellectual and organizational rather than physical. Some organizations define and follow a precise and detailed development process, while others may not even be able to describe their processes. Furthermore, every organization employs a process at least slightly different from that of every other organization. In fact, the same enterprise may follow different processes for each of several different types of development projects.

A well-defined development process is useful for the following reasons:

- *Quality assurance:* A development process specifies the phases a development project will pass through and the checkpoints along the way. When these phases and checkpoints are chosen wisely, following the development process is one way of assuring the quality of the resulting product.
- *Coordination:* A clearly articulated development process acts as a master plan which defines the roles of each of the players on the development team. This plan informs the members of the team when their contributions will be needed and with whom they will need to exchange information and materials.
- *Planning:* A development process contains natural milestones corresponding to the completion of each phase. The timing of these milestones anchors the schedule of the overall development project.
- *Management:* A development process is a benchmark for assessing the performance of an ongoing development effort. By comparing the actual events to the es-

tablished process, a manager can identify possible problem areas.
- *Improvement:* The careful documentation of an organization's development process often helps to identify opportunities for improvement.

The generic product development process consists of six phases, as illustrated in Figure 2.2. The process begins with a planning phase, which is the link to advanced research and technology development activities. The output of the planning phase is the project's mission statement, which is the input required to begin the concept development phase and which serves as a guide to the development team. The conclusion of the product development process is the product launch, at which time the product becomes available for purchase in the marketplace.

One way to think about the development process is as the initial creation of a wide set of alternative product concepts and then the subsequent narrowing of alternatives and increasing specification of the product until the product can be reliably and repeatably produced by the production system. Note that most of the phases of development are defined in terms of the state of the product, although the production process and marketing plans, among other tangible outputs, are also evolving as development progresses.

Another way to think about the development process is as an information-processing system. The process begins with inputs such as the corporate objectives and the capabilities of available technologies, product platforms, and production systems. Various activities process the development information, formulating specifications, concepts, and design details. The process concludes when all the information required to support production and sales has been created and communicated.

Figure 2.2 also identifies the key activities and responsibilities of the different functions of the organization during each development phase. Because of their continuous involvement in the process, we choose to articulate the roles of marketing, design, and manufacturing. Representatives from other functions, such as research, finance, field service, and sales, also play key roles at particular points in the process.

The six phases of the generic development process are:

0. *Planning:* The planning activity is often referred to as "phase zero" since it precedes the project approval and launch of the actual product development process. This phase begins with corporate strategy and includes assessment of technology developments and market objectives. The output of

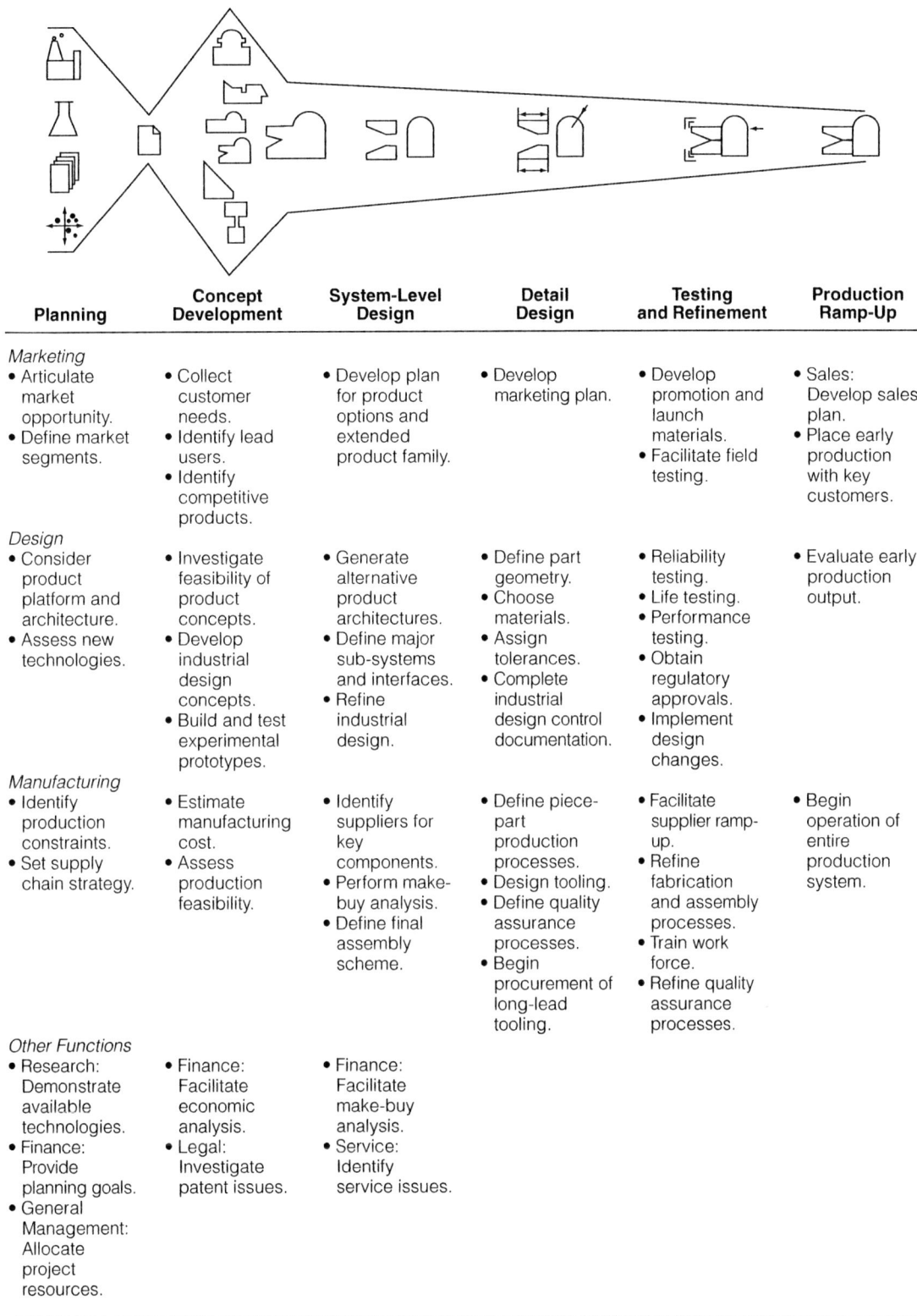

Planning	Concept Development	System-Level Design	Detail Design	Testing and Refinement	Production Ramp-Up
Marketing • Articulate market opportunity. • Define market segments.	• Collect customer needs. • Identify lead users. • Identify competitive products.	• Develop plan for product options and extended product family.	• Develop marketing plan.	• Develop promotion and launch materials. • Facilitate field testing.	• Sales: Develop sales plan. • Place early production with key customers.
Design • Consider product platform and architecture. • Assess new technologies.	• Investigate feasibility of product concepts. • Develop industrial design concepts. • Build and test experimental prototypes.	• Generate alternative product architectures. • Define major sub-systems and interfaces. • Refine industrial design.	• Define part geometry. • Choose materials. • Assign tolerances. • Complete industrial design control documentation.	• Reliability testing. • Life testing. • Performance testing. • Obtain regulatory approvals. • Implement design changes.	• Evaluate early production output.
Manufacturing • Identify production constraints. • Set supply chain strategy.	• Estimate manufacturing cost. • Assess production feasibility.	• Identify suppliers for key components. • Perform make-buy analysis. • Define final assembly scheme.	• Define piece-part production processes. • Design tooling. • Define quality assurance processes. • Begin procurement of long-lead tooling.	• Facilitate supplier ramp-up. • Refine fabrication and assembly processes. • Train work force. • Refine quality assurance processes.	• Begin operation of entire production system.
Other Functions • Research: Demonstrate available technologies. • Finance: Provide planning goals. • General Management: Allocate project resources.	• Finance: Facilitate economic analysis. • Legal: Investigate patent issues.	• Finance: Facilitate make-buy analysis. • Service: Identify service issues.			

Figure 2.2

The generic product development process. Six phases are shown, including the tasks and responsibilities of the key functions of the organization for each phase.

the planning phase is the project mission statement, which specifies the target market for the product, business goals, key assumptions, and constraints. Chapter 3, Product Planning, presents a discussion of this planning process.

1. *Concept development:* In the concept development phase, the needs of the target market are identified, alternative product concepts are generated and evaluated, and one or more concepts are selected for further development and testing. A concept is a description of the form, function, and features of a product and is usually accompanied by a set of specifications, an analysis of competitive products, and an economic justification of the project. This book presents several detailed methods for the concept development phase (Chapters 4-8). We expand this phase into each of its constitutive activities in the next section.

2. *System-level design:* The system-level design phase includes the definition of the product architecture and the decomposition of the product into subsystems and components. The final assembly scheme for the production system is usually defined during this phase as well. The output of this phase usually includes a geometric layout of the product, a functional specification of each of the product's subsystems, and a preliminary process flow diagram for the final assembly process. Chapter 9, Product Architecture, discusses some of the important activities of system-level design.

3. *Detail design:* The detail design phase includes the complete specification of the geometry, materials, and tolerances of all of the unique parts in the product and the identification of all of the standard parts to be purchased from suppliers. A process plan is established and tooling is designed for each part to be fabricated within the production system. The output of this phase is the *control documentation* for the product—the drawings or computer files describing the geometry of each part and its production tooling, the specifications of the purchased parts, and the process plans for the fabrication and assembly of the product. Chapter 11, Design for Manufacturing, provides a discussion of a few of the issues faced in the detail design phase.

4. *Testing and refinement:* The testing and refinement phase involves the construction and evaluation of multiple preproduction versions of the product. Early (*alpha*) prototypes are usually built with *production-intent* parts—parts with the same geometry and material properties as intended for the production version of the product but not necessarily fabricated with the actual processes to be used in production. Alpha prototypes are tested to determine whether or not the product will work as designed and whether or not the product satisfies the key customer needs. Later (*beta*) prototypes are usually built with parts supplied by the intended production processes but may not be assembled using the intended final assembly process. Beta prototypes are extensively evaluated internally and are also typically tested by customers in their own use environment. The goal for the beta prototypes is usually to answer questions about performance and reliability in order to identify necessary engineering changes for the final product. Chapter 12, Prototyping, presents a thorough discussion of the nature and use of prototypes.

5. *Production ramp-up:* In the production ramp-up phase, the product is made using the intended production system. The purpose of the ramp-up is to train the work force and to work out any remaining problems in the production processes. Products produced during production ramp-up are sometimes supplied to preferred customers and are carefully evaluated to identify any remaining flaws. The transition from production ramp-up to ongoing production is usually gradual. At some point in this transition, the product is *launched* and becomes available for widespread distribution.

2.2 | CONCEPT DEVELOPMENT: THE FRONT-END PROCESS

Because the concept development phase of the development process demands perhaps more coordination among functions than any other, many of the integrative development methods presented in this book are concentrated here. In this section we expand the concept development phase into what we call the *front-end process*. The front-end process generally contains many interrelated activities, ordered roughly as shown in Figure 2.3.

Rarely does the entire process proceed in purely sequential fashion, completing each activity before beginning the next. In practice, the front-end activities may be overlapped in time and iteration is often necessary. The dashed arrows in Figure 2.3 reflect the uncertain nature of progress in product development. At almost any stage, new information may become available or results

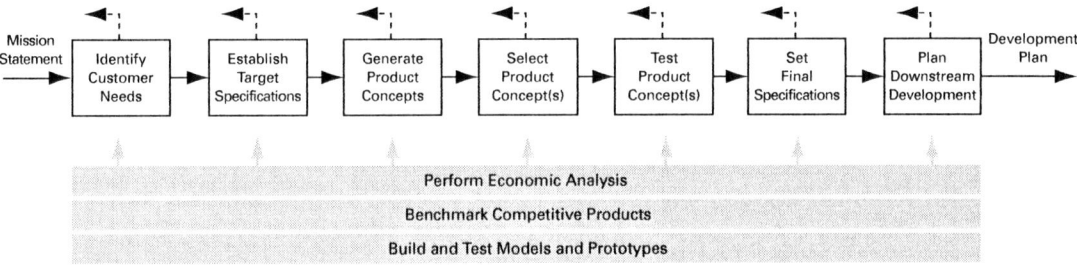

Figure 2.3
The many front-end activities comprising the concept development phase.

learned which can cause the team to step back to repeat an earlier activity before proceeding. This repetition of nominally complete activities is known as development *iteration.*

The concept development process includes the following activities:

- *Identifying customer needs:* The goal of this activity is to understand customers' needs and to effectively communicate them to the development team. The output of this step is a set of carefully constructed customer need statements, organized in a hierarchical list, with importance weightings for each need. A method for this activity is presented in Chapter 4, Identifying Customer Needs.
- *Establishing target specifications:* Specifications provide a precise description of what a product has to do. They are the translation of the customer needs into technical terms. Targets for the specifications are set early in the process and represent the hopes of the development team. Later these specifications are refined to be consistent with the constraints imposed by the team's choice of a product concept. The output of this stage is a list of target specifications. Each specification consists of a metric, and marginal and ideal values for that metric. A method for the specification activity is given in Chapter 5, Product Specifications.
- *Concept generation:* The goal of concept generation is to thoroughly explore the space of product concepts that may address the customer needs. Concept generation includes a mix of external search, creative problem solving within the team, and systematic exploration of the various solution fragments the team generates. The result of this activity is usually a set of 10 to 20 concepts, each typically represented by a sketch and brief descriptive text. Chapter 6, Concept Generation, describes this activity in detail.
- *Concept selection:* Concept selection is the activity in which various product concepts are analyzed and se-

quentially eliminated to identify the most promising concept(s). The process usually requires several iterations and may initiate additional concept generation and refinement. A method for this activity is described in Chapter 7, Concept Selection.

- *Concept testing:* One or more concepts are then tested to verify that the customer needs have been met, assess the market potential of the product, and identify any shortcomings which must be remedied during further development. If the customer response is poor, the development project may be terminated or some earlier activities may be repeated as necessary. Chapter 8, Concept Testing, explains a method for this activity.
- *Setting final specifications:* The target specifications set earlier in the process are revisited after a concept has been selected and tested. At this point, the team must commit to specific values of the metrics reflecting the constraints inherent in the product concept, limitations identified through technical modeling, and trade-offs between cost and performance. Chapter 5, Product Specifications, explains the details of this activity.
- *Project planning:* In this final activity of concept development, the team creates a detailed development schedule, devises a strategy to minimize development time, and identifies the resources required to complete the project. The major results of the front-end activities can be usefully captured in a *contract book* which contains the mission statement, the customer needs, the details of the selected concept, the product specifications, the economic analysis of the product, the development schedule, the project staffing, and the budget. The contract book serves to document the agreement (contract) between the team and the senior management of the enterprise. A project planning method is presented in Chapter 14, Managing Projects.
- *Economic analysis:* The team, often with the support of a financial analyst, builds an economic model for the

	Generic (Market Pull)	Technology Push	Platform Products	Process Intensive	Customized
Description	The firm begins with a market opportunity, then finds appropriate technologies to meet customer needs.	The firm begins with a new technology, then finds an appropriate market.	The firm assumes that the new product will be built around an established technological subsystem.	Characteristics of the product are highly constrained by the production process.	New products are slight variations of existing configurations.
Distinctions with respect to generic process		Planning phase involves matching technology and market. Concept development assumes a given technology.	Concept development assumes a technology platform.	Both process and product must be developed together from the very beginning, or an existing production process must be specified from the beginning.	Similarity of projects allows for a highly structured development process.
Examples	Most sporting goods, furniture, tools.	Gore-Tex rainwear, Tyvek envelopes.	Consumer electronics, computers, printers.	Snack foods, cereal, chemicals, semiconductors.	Switches, motors, batteries, containers.

Figure 2.4
Summary of variants of generic development process.

new product. This model is used to justify continuation of the overall development program and to resolve specific trade-offs among, for example, development costs and manufacturing costs. Economic analysis is shown as one of the ongoing activities in the concept development phase. An early economic analysis will almost always be performed before the project even begins, and this analysis is updated as more information becomes available. A method for this activity is presented in Chapter 13, Product Development Economics.

- *Benchmarking of competitive products:* An understanding of competitive products is critical to successful positioning of a new product and can provide a rich source of ideas for the product and production process design. Competitive *benchmarking* is performed in support of many of the front-end activities. Various aspects of competitive benchmarking are presented in Chapters 4-8.
- *Modeling and prototyping:* Every stage of the concept development process involves various forms of models and prototypes. These may include, among others: early "proof-of-concept" models, which help the development team to demonstrate feasibility; "form-only" models, which can be shown to customers to evaluate ergonomics and style; and spreadsheet models of technical trade-offs. Methods for modeling and prototyping are discussed throughout the book, including in Chapters 4, 5, 6, 8, 10, and 12.

2.3 ADAPTING THE GENERIC PRODUCT DEVELOPMENT PROCESS

The development process described by Figures 2.2 and 2.3 is generic, and particular processes will differ in accordance with a firm's unique context. The generic process is most like the process used in a *market-pull* situation: a firm begins product development with a market opportunity and then uses whatever available technologies are required to satisfy the market need (i.e., the market "pulls" the development decisions). In addition to the market-pull process outlined in Figures 2.2 and 2.3, several variants are common and correspond to the following: *technology-push* products, *platform* products, *process-intensive* products, and *customized* products. Each of these situations is described below. The characteristics of these situations and the resulting deviations from the generic process are summarized in Figure 2.4.

Technology-Push Products

In developing technology-push products, the firm begins with a new proprietary technology and looks for an appropriate market in which to apply this technology (that is, the technology "pushes" development). Gore-Tex, an expanded Teflon sheet manufactured by W.L. Gore Associates, is a striking example of technology push. The company has developed dozens of products incorporating Gore-Tex, including artificial veins for vascular surgery, insulation for high-performance electric cables, fabric for outerwear, dental floss, and liners for bagpipe bags.

Many successful technology-push products involve basic materials or basic process technologies. This may be because basic materials and processes are deployed in thousands of applications, and there is therefore a high likelihood that new and unusual characteristics of materials and processes can be matched with an appropriate application.

The generic product development process can be used with minor modifications for technology-push products. The technology-push process begins with the planning phase, in which the given technology is matched with a market opportunity. Once this matching has occurred, the remainder of the generic development process can be followed. The team includes an assumption in the mission statement that the particular technology will be embodied in the product concepts considered by the team. Although many extremely successful products have arisen from technology-push development, this approach can be perilous. The product is unlikely to succeed unless (1)the assumed technology offers a clear competitive advantage in meeting customer needs, and (2)suitable alternative technologies are unavailable or very difficult for competitors to utilize. Project risk can possibly be minimized by simultaneously considering the merit of a broader set of concepts which do not necessarily incorporate the new technology. In this way the team verifies that the product concept embodying the new technology is superior to the alternatives.

Platform Products

A platform product is built around a preexisting technological subsystem (a technology *platform*). Examples of such platforms include the tape transport mechanism in the Sony Walkman, the Apple Macintosh operating system, and the instant film used in Polaroid cameras. Huge investments were made in developing these platforms, and therefore every attempt is made to incorporate them into several different products. In some sense, platform products are very similar to technology-push products in that the team begins the development effort with an assumption that the product concept will embody a particular technology. The primary difference is that a technology platform has already demonstrated its usefulness in the marketplace in meeting customer needs. The firm can in many cases assume that the technology will also be useful in related markets. Products built on technology platforms are much simpler to develop than if the technology were developed from scratch. For this reason, and because of the possible sharing of costs across several products, a firm may be able to

offer a platform product in markets that could not justify the development of a unique technology.

Process-Intensive Products

Examples of process-intensive products include semiconductors, foods, chemicals, and paper. For these products, the production process places strict constraints on the properties of the product, so that the product design cannot be separated, even at the concept phase, from the production process design. In many cases, process-intensive products are produced in very high volumes and are bulk, as opposed to discrete, goods.

In some situations, a new product and new process are developed simultaneously. For example, creating a new shape of breakfast cereal or snack food will require both product and process development activities. In other cases, a specific existing process for making the product is chosen in advance, and the product design is constrained by the capabilities of this process. This might be true of a new paper product to be made in a particular paper mill or a new semiconductor device to be made in an existing wafer fabrication facility.

Customized Products

Examples of customized products include switches, motors, batteries, and containers. Customized products are slight variations of standard configurations and are typically developed in response to a specific order by a customer. Development of customized products consists primarily of setting values of design variables such as physical dimensions and materials. When a customer requests a new product, the firm executes a structured design and development process to create the product to meet the customer's needs. Such firms typically have created a highly detailed development process involving a well-defined sequence of steps with a structured flow of information (analogous to a production process). For customized products, the generic process is augmented with a detailed description of the specific information-processing activities required within each of the phases. Such development processes may consist of hundreds of carefully defined activities.

2.4 | THE AMF DEVELOPMENT PROCESS

AMF Bowling is a market-pull enterprise. AMF generally drives its development process with a market need and seeks out whatever technology is required to meet that need. Its competitive advantage arises from strong marketing channels, strong brand recognition, and a large installed base of equipment, not from any single proprietary technology. For this reason, the technology-push approach would not be appropriate. AMF products are assembled from components fabricated with relatively conventional processes such as molding, casting, and machining. So the AMF product is clearly not process intensive in the way a food product or a chemical is. Bowling equipment is rarely customized for a particular customer; most of the product

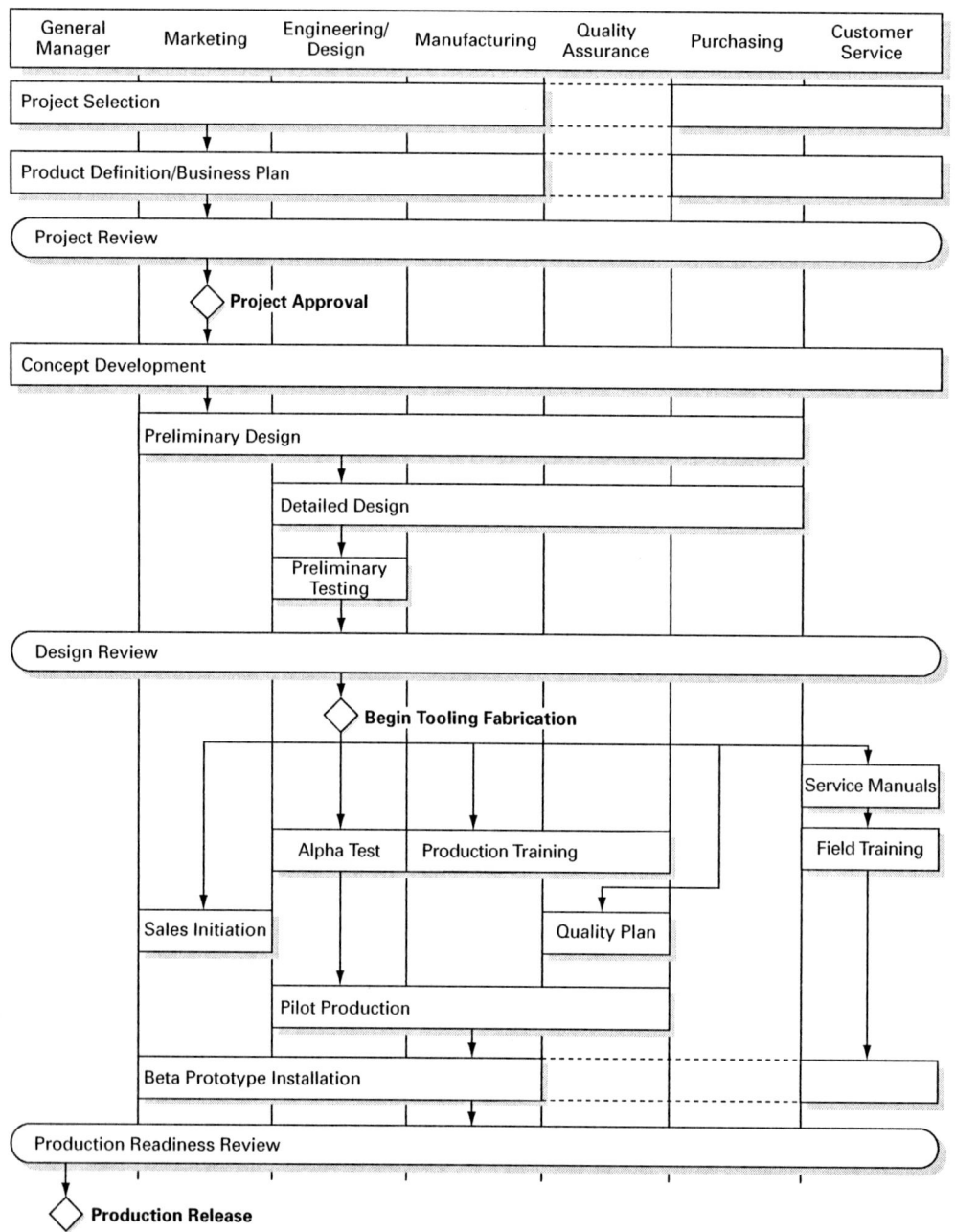

Figure 2.5
AMF Bowling's standard development process.

development at AMF is aimed at new models of products, rather than at the customization of existing models. For this reason, the customization approach is also inappropriate.

AMF chose to establish a development process similar to the generic process. The process proposed by the AMF engineering manager is illustrated in Figure 2.5. The representation of the development process used by AMF is a hybrid of those used in Figures 2.2 and 2.3, in that it shows the individual activities in the development process as well as the roles of the different development functions in those activities. Note that AMF defines the key functions in product development as marketing, engineering/design, manufacturing, quality assurance, purchasing, and customer service. Also note that there are three major milestones in the process: the project approval, the beginning of tooling fabrication, and the production release. Each of these milestones follows a major review.

Although AMF established a standard process, its managers realized that this process would not necessarily be suitable in its entirety for all AMF products. For example, a few of AMF's new products are based on technology platforms. When platform products are developed, the team assumes the use of an existing technology platform during concept development. Nevertheless, the standard development process is the baseline from which a particular project plan begins.

2.5 | PRODUCT DEVELOPMENT ORGANIZATIONS

In addition to crafting an effective development process, successful firms must organize their product development staffs effectively. In this section, we describe several types of organizations used for product development and offer guidelines for choosing among these options.

Organizations Are Formed by Establishing Links among Individuals

A product development organization is the scheme by which individual designers and developers are linked together into groups. The links among individuals may be formal or informal and include, among others, these types:

- *Reporting relationships:* Reporting relationships give rise to the classic notion of *supervisor* and *subordinate.* These are the formal links most frequently shown on an organization chart.
- *Financial arrangements:* Individuals are linked by being part of the same financial entity, such as that de-

fined by a particular budget category or profit-and-loss statement.
- *Physical layout:* Links are created between individuals when they share the same office, floor, building, or site. These links are often informal, arising from spontaneous encounters while at work.

Any particular individual may be linked in several different ways to other individuals. For example, an engineer may be linked by a reporting relationship to another engineer in a different building, while being linked by physical layout to a marketing person sitting in the next office. The strongest organizational links are typically those involving performance evaluation, budgets, and other resource allocations.

Organizational Links May Be Aligned with Functions, Projects, or Both

Regardless of their organizational links, particular individuals can be classified in two different ways: according to their *function* and according to the *projects* they work on.

- A function (in organizational terms) is an area of responsibility usually involving specialized education, training, or experience. The classic functions in product development organizations are marketing, design, and manufacturing. Finer divisions than these are also possible and may include, for example, market research, market strategy, stress analysis, industrial design, human factors engineering, process development, and operations management.
- Regardless of their functions, individuals apply their expertise to specific projects. In product development, a project is the set of activities in the development process for a particular product and includes, for example, identifying customer needs, generating product concepts, etc.

Note that these two classifications must overlap: individuals from several different functions will work on the same project. Also, while most individuals are associated with only one function, they may contribute to more than one project. Two classic organizational structures arise from aligning the organizational links according to function or according to projects. In *functional organizations,* the organizational links are primarily among those who perform similar functions. In *project organizations,* the organizational links are primarily among those who work on the same project.

For example, a strict functional organization might include a group of marketing professionals, all sharing similar training and expertise. These people would all report to the same manager, who would evaluate them and set their salaries. The group would have its own budget and the people would sit in the same part of a building. This marketing group would be involved in many different projects, but there would be no strong organizational links to the other members of each project team. There would be similarly arranged groups corresponding to design and to manufacturing.

A strict project organization would be made up of groups of people from several different functions, with each group focused on the development of a specific product (or product line). These groups would each report to an experienced project manager, who might be drawn from any of the functional areas. Performance evaluation would be handled by the project manager, and members of the team would typically be colocated as much as possible so that they all work in the same office or part of a building. New ventures, or "start-ups," are among the most extreme examples of project organizations: every individual, regardless of function, is linked together by a single project—the growth of the new company and the creation of its product(s). In these settings, the president or CEO can be viewed as the project manager. Established firms will sometimes form a "tiger team" with dedicated resources for a single project when special focus is required to complete an important development project.

The *matrix organization* was conceived as a hybrid of functional and project organizations. In the matrix organization, individuals are linked to others according to both the project they work on and their function. Typically each individual has two supervisors, one a project manager and one a functional manager. The practical reality is that either the project or the function tends to have stronger links. This is because, for example, both functional and project managers cannot have independent budget authority, they cannot independently evaluate and determine the salaries of their subordinates, and both functional and project organizations cannot easily be grouped together physically. As a result, either the functional or the project organization tends to dominate.

Two variants of the matrix organization are called the *heavyweight project organization* and *lightweight project organization* (Hayes et al., 1988). A heavyweight project organization contains strong project links. The heavyweight project manager has complete budget authority, is heavily involved in performance evaluation of the team members, and makes most of the major resource allocation decisions. Although each participant in a project also belongs to a functional organization, the functional managers have relatively little authority and control. A heavyweight project team in various industries may be called an *integrated product team* (IPT), a *design-build team* (DBT), or simply a *product development team* (PDT). Each of these terms emphasizes the cross-functional nature of these teams.

A lightweight project organization contains weaker project links and relatively stronger functional links. In this scheme, the project manager is more of a coordinator and administrator. The lightweight project manager updates schedules, arranges meetings, and facilitates coordination, but the manager has no real authority and control in the project organization. The functional managers are responsible for budgets, hiring and firing, and performance evaluation. Figure 2.6 illustrates the pure functional and project organizations, along with the heavyweight and lightweight variants of the matrix organization.

In this book we refer to the *project team* as the primary organizational unit. In this context, the team is the set of all people involved in the project, regardless of the organizational structure of the product development staff. In a functional organization, the team consists of individuals distributed throughout the functional groups without any organizational linkages other than their common involvement in a project. In the other organizations, the team corresponds to a formal organizational entity, the project group, and has a formally appointed manager. For this reason the notion of a team has much more meaning in matrix and project organizations than it does in functional organizations.

Choosing an Organizational Structure

The most appropriate choice of organizational structure depends on which organizational performance factors are most critical to success. Functional organizations tend to breed specialization and deep expertise in the functional areas. Project organizations tend to enable rapid and effective coordination among diverse functions. Matrix organizations, being hybrids, have the potential to exhibit some of each of these characteristics. The following questions help guide the choice of organizational structure:

- *How important is cross-functional integration?* Functional organizations may exhibit difficulty in coordinating project decisions which span the functional areas. Project organizations tend to enable strong cross-functional integration because of the organizational links of the team members across the functions.
- *How critical is cutting-edge functional expertise to business success?* When disciplinary expertise must be developed and retained over several product generations, then some functional links are necessary. For example, in some aerospace companies, computational

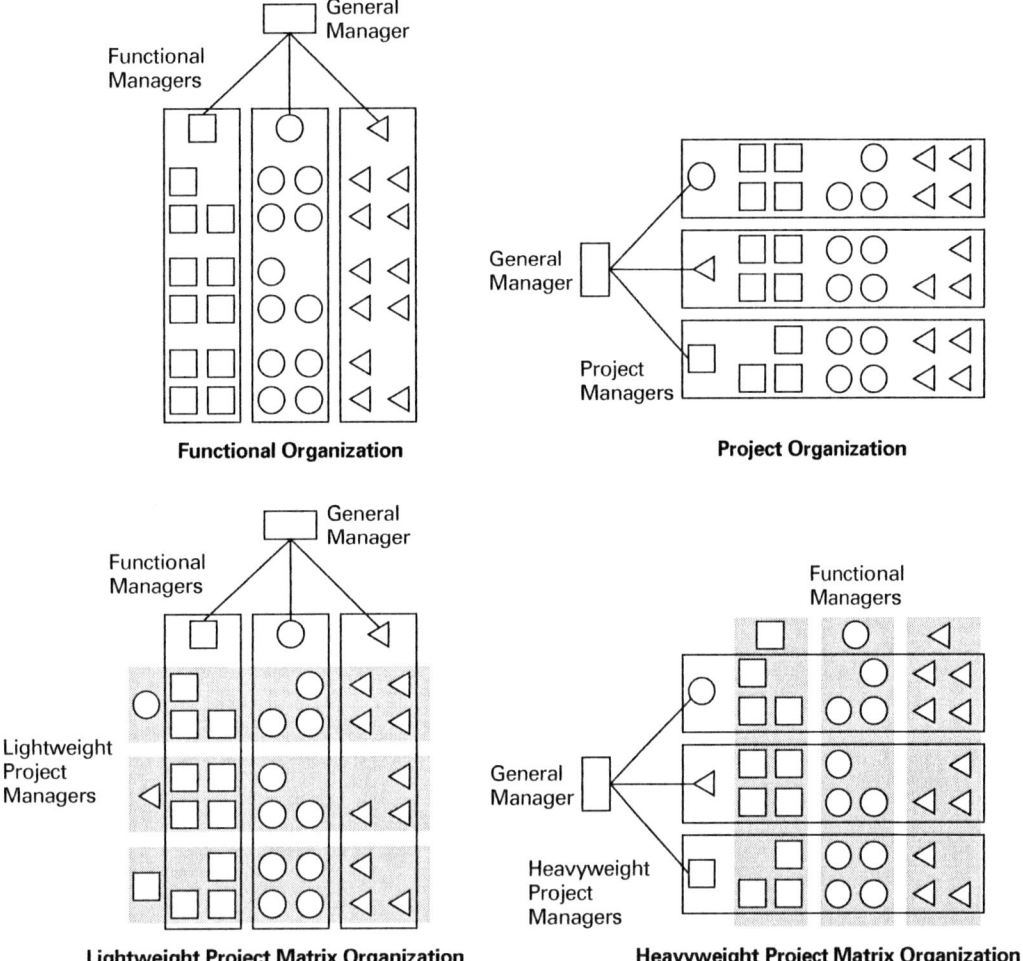

Figure 2.6
Various product development organizations. For simplicity, three functions and three projects are shown.
(Adapted from Hayes, et al., 1988.)

fluid dynamics is so critical that the fluid dynamicists are organized functionally to ensure the firm will have the best possible capability in this area.

- *Can individuals from each function be fully utilized for most of the duration of a project?* For example, a project may only require a portion of an industrial designer's time for a fraction of the duration of a project. In order to use industrial design resources efficiently, the firm may choose to organize the industrial designers functionally, so that several projects can draw on the industrial design resource in exactly the amount needed for a particular project.
- *How important is product development speed?* Project organizations tend to allow for conflicts to be resolved

quickly and for individuals from different functions to coordinate their activities efficiently. Relatively little time is spent transferring information, assigning responsibilities, and coordinating tasks. For this reason, project organizations are usually faster than functional organizations in developing innovative products. For example, portable computer manufacturers almost always organize their product development teams by project. This allows the teams to develop new products within the extremely short periods required by the fast-paced computer market.

Dozens of other issues confound the choice between functional and project organizations. Figure 2.7 summa-

	Functional Organization	Matrix Organization		Project Organization
		Lightweight Project Organization	Heavyweight Project Organization	
Strengths	Fosters development of deep specialization and expertise.	Coordination and administration of projects is explicitly assigned to a single project manager. Maintains development of specialization and expertise.	Provides integration and speed benefits of the project organization. Some of the specialization of a functional organization is retained.	Resources can be optimally allocated within the project team. Technical and market trade-offs can be evaluated quickly.
Weaknesses	Coordination among different functional groups can be slow and bureaucratic.	Requires more managers and administrators than a non-matrix organization.	Requires more managers and administrators than a non-matrix organization.	Individuals may have difficulty maintaining cutting-edge functional capabilities.
Typical examples	Customization development—firms in which development involves slight variations to a standard design (e.g., custom motors, bearings, packaging).	Traditional automobile, electronics, and aerospace companies.	Many recently successful projects in automobile, electronics, and aerospace companies.	Start-up companies. "Tiger teams" and "skunk works" intended to achieve breakthroughs. Firms competing in extremely dynamic markets.
Major issues	How to integrate different functions (e.g., marketing and design) to achieve a common goal.	How to balance functions and projects. How to simultaneously evaluate project and functional performance.		How to maintain functional expertise over time. How to share technical learning from one project to another.

Figure 2.7
Characteristics of different organizational structures.

rizes some of the strengths and weaknesses of each organizational type, examples of the types of firms pursuing each strategy, and the major issues associated with each approach.

2.6 | THE AMF ORGANIZATION

AMF chose to organize its product development staff in a matrix structure. The functions involved in product development at AMF include engineering, manufacturing, marketing, sales, purchasing, and quality assurance. Each of these functions has a manager who reports to the general manager of the division. However, product development projects are led by project managers and project teams are drawn from each of the functional areas. The AMF matrix organization is probably closest to the lightweight project organization. This is because the project managers are not typically the most senior managers in the division and do not have direct control of resources and staffing for the project teams. While in general a lightweight project organization tends to strengthen the functions at the expense of project efficiency, several characteristics of the AMF organization make the lightweight organization a wise choice and have led to good product development performance.

The most significant factor leading to the choice of a lightweight project organization is that AMF carries out many small product development projects along with one or two large projects. The result of this mix of projects is that many of the team members on smaller projects contribute on a part-time basis. By having relatively strong functional links between individuals, the assignment of staff to smaller projects and the balancing of workload within a function are more easily accomplished.

Another factor allowing AMF to use a lightweight project organization and still achieve high performance in product development is that AMF is an extraordinarily lean company. The Capital Equipment Division has fewer than 100 salaried employees generating and supporting sales of over $100 million per year. Everyone in the division works in the same building, and most of the key employees earn substantial financial rewards when the division is highly profitable. As a result, members of project teams are motivated to look beyond their own functions and work together to develop successful products.

A slight deviation from the standard lightweight project organization also facilitates project completion. The engineering manager is held personally responsible for all aspects of successful completion of projects and not for engineering excellence alone. Although he is responsible for the engineering function, he is primarily responsible for developing successful products. He therefore works daily to ensure that the appropriate coordination occurs, for example, between marketing and engineering.

Finally, the emphasis that the senior management places on product development encourages effective teamwork. The general manager takes a personal interest in every product development project and devotes several days each month to monitoring the progress of these projects. The message communicated to the project teams is that successful products are more important than strong functions.

2.7 | SUMMARY

An enterprise must make two important decisions about the way it carries out product development. It must define both a product development process and a product development organization.

- A product development process is the sequence of steps an enterprise employs to conceive, design, and commercialize a product.
- A well-defined development process helps to assure product quality, facilitate coordination among team members, plan the development project, and continuously improve the process.
- The generic process presented in this chapter includes six phases: planning, concept development, system-level design, detail design, testing and refinement, and production ramp-up.
- The concept development phase requires tremendous integration across the different functions on the development team. This front-end process includes identifying customer needs, analyzing competitive products, establishing target specifications, generating product concepts, selecting one or more final concepts, setting final specifications, testing the concept(s), performing

an economic analysis, and planning the remaining project activities. The results of the concept development phase are documented in a contract book.
- The development process employed by a particular firm may differ somewhat from the generic process described here. The generic process is most appropriate for market-pull products. Other types of products, which may require variants of the generic process, include technology-push products, platform products, process-intensive products, and customized products.
- Regardless of the development process, tasks are completed by individuals residing in organizations. Organizations are defined by linking individuals through reporting relationships, financial relationships, and/or physical layout.
- Functional organizations are those in which the organizational links correspond to the development functions. Project organizations are those in which the organizational links correspond to the development projects. Two types of hybrid, or matrix, organizations are the heavyweight project organization and the lightweight project organization.
- The classic trade-off between functional organizations and project organizations is between deep functional expertise and coordination efficiency.

2.8 | REFERENCES AND BIBLIOGRAPHY

Many current resources are available on the Internet via

www.ulrich-eppinger.net

The concept of heavyweight and lightweight project organizations is articulated by Hayes, Wheelwright, and Clark. Wheelwright and Clark also discuss product strategy, planning, and technology development activities which generally precede the product development process.

Hayes, Robert H., Steven C. Wheelwright, and Kim B. Clark, *Dynamic Manufacturing: Creating the Learning Organization,* The Free Press, New York, 1988.

Wheelwright, Steven C., and Kim B. Clark, *Revolutionizing Product Development: Quantum Leaps in Speed, Efficiency, and Quality,* The Free Press, New York, 1992.

Andreasen and Hein provide some good ideas on how to integrate different functions in product development. They also show several conceptual models of product development organizations.

Andreasen, M. Myrup, and Lars Hein, *Integrated Product Development,* Springer-Verlag, New York, 1987.

Allen provides strong empirical evidence that physical layout can be used to create significant, although informal, organizational links. He also discusses the use of matrix organizations to mitigate the weaknesses of functional and project organizations.

Allen, Thomas J., *Managing the Flow of Technology: Technology Transfer and the Dissemination of Technological Information within the R&D Organization,* MIT Press, Cambridge, MA, 1977.

Galbraith's seminal book on organizational design contains much useful information which can be applied to product development. His 1994 book is an update of his earlier writing.

Galbraith, Jay R., *Designing Complex Organizations,* Addison-Wesley, Reading, MA, 1973.

Galbraith, Jay R., *Competing with Flexible Lateral Organizations,* second edition, Addison-Wesley, Reading, MA, 1994.

Exercises

1. Diagram a process for planning and cooking a family dinner. Does your process resemble the generic product development process? Is cooking dinner analogous to a market-pull, technology-push, process-intensive, or customization process?

2. Define a process for finding a job. For what types of endeavors does a well-defined process enhance performance?

3. What type of development process would you expect to find in an established company successful at developing residential air-conditioning units? How about for a small company that is trying to break into the market for racing wheelchairs?

4. Sketch the organization (in some appropriate graphical representation) of a consulting firm that develops new products for clients on a project-by-project basis. Assume that the individuals in the firm represent all of the different functions required to develop a new product. Would this organization most likely be aligned with functions, be aligned by projects, or be a hybrid?

Thought Questions

1. What role does basic technological research play in the product development process? How would you modify Figure 2.3 to better represent the research and technology development activities in product development?

2. Is there an analogy between a university and a product development organization? Is a university a functional or project organization?

3. What is the product development organization for students engaged in projects as part of a product development class?

4. Is it possible for some members of a product development organization to be organized functionally, while others are organized by project? If so, which members of the team would be the most likely candidates for the functional organization?

Identifying Customer Needs 3

A successful hand tool manufacturer was exploring the growing market for hand-held power tools. After performing initial research, the firm decided to enter the market with a cordless screwdriver. Figure 3.1 shows several existing products used to drive screws. After some initial concept work, the manufacturer's development team fabricated and field-tested several prototypes. The results were discouraging. Although some of the products were liked better than others, each one had some feature that customers objected to in one way or another. The results were quite mystifying since the company had been successful in related consumer products for years. After much discussion, the team decided that its process for identifying customer needs was inadequate.

This chapter presents a method for comprehensively identifying a set of customer needs. The goals of the method are to:

- Ensure that the product is focused on customer needs.
- Identify latent or hidden needs as well as explicit needs.
- Provide a fact base for justifying the product specifications.
- Create an archival record of the needs activity of the development process.
- Ensure that no critical customer need is missed or forgotten.
- Develop a common understanding of customer needs among members of the development team.

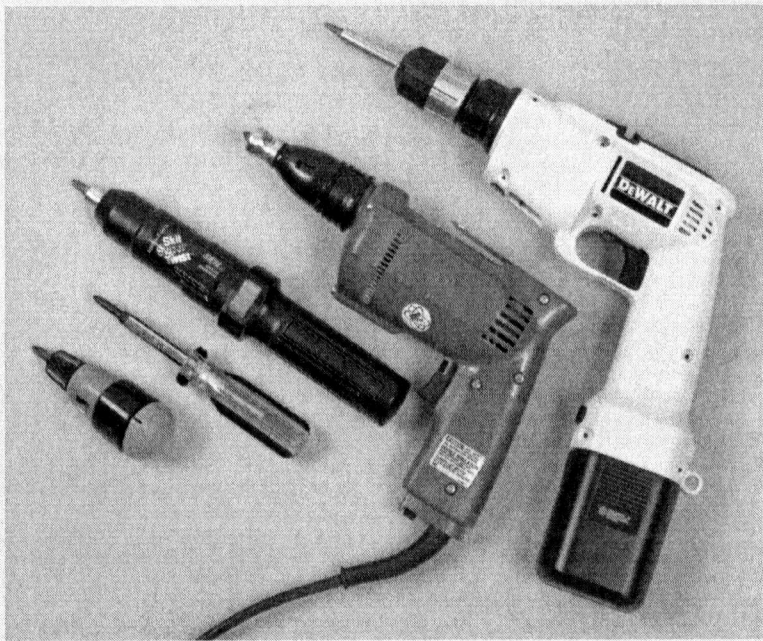

Figure 3.1
Existing products used to drive screws: manual screwdrivers, cordless screwdriver, screw gun, cordless drill with driver bit. Stuart Cohen

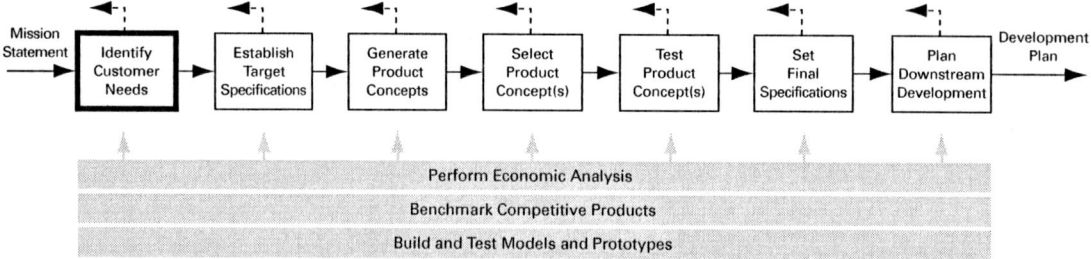

Figure 3.2
The customer-needs activity in relation to other concept development activities.

The philosophy behind the method is to create a high-quality information channel that runs directly between customers in the target market and the developers of the product. This philosophy is built on the premise that those who directly control the details of the product, including the engineers and industrial designers, must interact with customers and experience the *use environment* of the product. Without this direct experience, technical trade-offs are not likely to be made correctly, innovative solutions to customer needs may never be discovered, and the development team may never develop a deep commitment to meeting customer needs.

The process of identifying customer needs is an integral part of the larger product development process and is most closely related to concept generation, concept selection, competitive benchmarking, and the establishment of product specifications. The customer-needs activity is shown in Figure 3.2 in relation to these other front-end product development activities, which collectively can be thought of as the *concept development* phase.

The concept development process illustrated in Figure 3.2 implies a distinction between customer needs and product specifications. This distinction is subtle but important. *Needs* are largely independent of any particular product we might develop; they are not specific to the concept we eventually choose to pursue. A team should be able to identify customer needs without knowing if or how it will eventually address those needs. On the other hand, *specifications* do depend on the concept we select. The specifications for the product we finally choose to develop will depend on what is technically and economically feasible and on what our competitors offer in the marketplace, as well as on customer needs. (See Chapter 5, Product Specifications, for a more detailed discussion of this distinction.) Also note that we choose to use the word *need* to label any attribute of a potential product that is desired by the customer; we do not distinguish here between a want and a need. Other terms used in industrial practice to refer to customer needs include *customer attributes* and *customer requirements*.

Identifying customer needs is itself a process, for which we present a five-step method. We believe that a little structure goes a long way in facilitating effective product development practices, and we hope and expect that this method will be viewed by those who employ it not as a rigid process but rather as a starting point for continuous improvement and refinement. The five steps are:

1. Gather raw data from customers.
2. Interpret the raw data in terms of customer needs.
3. Organize the needs into a hierarchy of primary, secondary, and (if necessary) tertiary needs.
4. Establish the relative importance of the needs.
5. Reflect on the results and the process.

We treat each of the five steps in turn and illustrate the key points with the cordless screwdriver example. We chose the screwdriver because it is simple enough that the method is not hidden by the complexity of the example. However, note that the same method, with minor adaptation, has been successfully applied to hundreds of products ranging from kitchen utensils costing less than $10 to machine tools costing hundreds of thousands of dollars.

Before beginning the development project, the firm typically specifies a particular market opportunity and lays out the broad constraints and objectives for the project. This information is frequently formalized as a *mission statement* (also sometimes called a *charter* or a *design brief*). The mission statement specifies which direction to go in but generally does not specify a precise destination or a particular way to proceed. The mission statement is the result of the product planning activities described in Chapter 3, Product Planning. The mission statement for the cordless screwdriver is shown in Figure 3.3.

The cordless screwdriver category of products is already relatively well developed. Such products are particularly well suited to a structured process for gathering customer needs. One could reasonably ask whether a structured method is effective for completely new categories of prod-

Mission Statement: Screwdriver Project

Product Description	• A hand-held, power-assisted device for installing threaded fasteners
Key Business Goals	• Product introduced in fourth quarter of 2002 • 50% gross margin • 10% share of cordless screwdriver market by 2004
Primary Market	• Do-it-yourself consumer
Secondary Markets	• Casual consumer • Light-duty professional
Assumptions	• Hand-held • Power-assisted • Nickel-metal-hydride rechargeable battery technology
Stakeholders	• User • Retailer • Sales force • Service center • Production • Legal department

Figure 3.3
Mission statement for the cordless screwdriver.

ucts with which customers have no experience. Satisfying needs is just as important in revolutionary products as in incremental products. A necessary condition for product success is that a product offer perceived benefits to the customer. Products offer benefits when they satisfy needs. This is true whether the product is an incremental variation on an existing product or whether it is a completely new product based on a revolutionary invention. Developing an entirely new category of product is a risky undertaking, and to some extent the only real indication of whether customer needs have been identified correctly is whether customers like the team's first prototypes. Nevertheless, in our opinion, a structured method for gathering data from customers remains useful and can lower the inherent risk in developing a radically new product. Whether or not customers are able to fully articulate their latent needs, interaction with customers in the target market will help the development team build a personal understanding of the user's environment and point of view. This information is always useful, even if it does not result in the identification of every need the new product will address.

3.1 | STEP 1: GATHER RAW DATA FROM CUSTOMERS

Consistent with our basic philosophy of creating a high-quality information channel directly from the customer,

gathering data involves contact with customers and experience with the use environment of the product. Three methods are commonly used:

1. *Interviews:* One or more development team members discuss needs with a single customer. Interviews are usually conducted in the customer's environment and typically last one to two hours.

2. *Focus groups:* A moderator facilitates a two-hour discussion with a group of 8 to 12 customers. Focus groups are typically conducted in a special room equipped with a two-way mirror allowing several members of the development team to observe the group. In most cases, the moderator is a professional market researcher, but a member of the development team sometimes moderates. The proceedings are usually videotaped. Participants are usually paid a modest fee ($50 to $100 each) for their attendance. The total cost of a focus group, including rental of the room, participant fees, videotaping, and refreshments is about $2,500. In most U.S. cities, firms that recruit participants, moderate focus groups and/or rent facilities are listed in the telephone book under "Market Research."

3. *Observing the product in use:* Watching customers use an existing product or perform a task for which a new product is intended can reveal important details about customer needs. For example,

a customer painting a house may use a screwdriver to open paint cans in addition to driving screws. Observation may be completely passive, without any direct interaction with the customer, or may involve working side by side with a customer, allowing members of the development team to develop firsthand experience using the product. Ideally, team members observe the product in the actual use environment. For some products, such as do-it-yourself tools, actually using the products is simple and natural; for others, such as surgical instruments, the team may have to use the products on surrogate tasks (e.g., cutting fruit instead of human tissue when developing a new scalpel).

Some practitioners also rely on written surveys for gathering raw data. While a mail survey is quite useful later in the process, we cannot recommend this approach for initial efforts to identify customer needs; written surveys simply do not provide enough information about the use environment of the product, and they are generally ineffective in revealing unanticipated needs.

Research by Griffin and Hauser shows that one 2-hour focus group reveals about the same number of needs as two 1-hour interviews (Griffin and Hauser, 1993). (See Figure 3.4.) Because interviews are usually less costly (per hour) than focus groups and because an interview often allows the product development team to experience the use environment of the product, we recommend that

interviews be the primary data collection method. Interviews may be supplemented with one or two focus groups as a way to allow top management to observe a group of customers or as a mechanism for sharing a common customer experience (via videotape) with the members of a larger team. Some practitioners believe that for certain products and customer groups, the interactions among the participants of focus groups can elicit more varied needs than are revealed through interviews, although this belief is not strongly supported by research findings.

Choosing Customers

Griffin and Hauser have also addressed the question of how many customers to interview in order to reveal most of the customer needs. In one study, they estimated that 90 percent of the customer needs for picnic coolers were revealed after 30 interviews. In another study, they estimated that 98 percent of the customer needs for a piece of office equipment were revealed after 25 hours of data collection in both focus groups and interviews. As a practical guideline for most products, conducting fewer than 10 interviews is probably inadequate and 50 interviews are probably too many. However, interviews can be conducted sequentially and the process can be terminated when no new needs are revealed by additional interviews. These guidelines apply to cases in which the development team is addressing a single market segment. If the team wishes to gather customer needs from multiple distinct segments, then the team may need to conduct 10 or more interviews in each segment. Concept development teams consisting of more than 10 people usually collect data from plenty of customers simply by involving much of the team in the process. For example, if a 10-person team is divided into five pairs and each pair conducts 6 interviews, the team conducts 30 interviews in total.

Needs can be identified more efficiently by interviewing a class of customers called *lead users*. According to von Hippel, lead users are customers who experience needs months or years ahead of the majority of the market and stand to benefit substantially from product innovations (von Hippel, 1988). These customers are particularly useful sources of data for two reasons: (1) they are often able to articulate their emerging needs, because they have had to struggle with the inadequacies of existing products, and (2) they may have already invented solutions to meet their needs. By focusing a portion of the data collection efforts on lead users, the team may be able to identify needs which, although explicit for lead users, are still latent for the majority of the market. Developing products to meet these latent needs allows a firm to anticipate trends and to leapfrog competitive products.

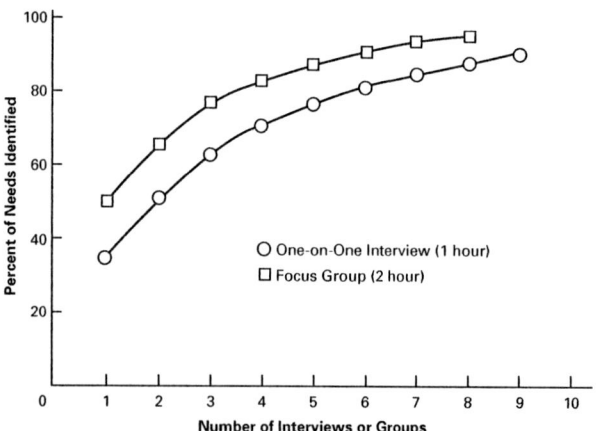

Figure 3.4

Comparison of the percentages of customer needs that are revealed for focus groups and interviews as a function of the number of sessions. Note that a focus group lasts two hours, while an interview lasts one hour.

(Source: Griffin and Hauser, 1993.)

	Lead Users	Users	Retailer or Sales Outlet	Service Centers
Homeowner (occasional use)	0	5	2	3
Handy person (frequent use)	3	10		
Professional (heavy-duty use)	3	2	2	

Figure 3.5
Customer selection matrix for the cordless screwdriver project.

The choice of which customers to interview is complicated when several different groups of people can be considered "the customer." For many products, one person (the buyer) makes the buying decision and another person (the user) actually uses the product. A good approach is to gather data from the end user of the product in all situations, and in cases where other types of customers and stakeholders are clearly important, to gather data from these people as well.

A customer selection matrix is useful for planning exploration of both market and customer variety. Burchill suggests that market segments be listed on the left side of the matrix while the different types of customers are listed across the top (Burchill et al., 1997), as shown in Figure 3.5. The number of intended customer contacts is entered in each cell to indicate the depth of coverage.

For industrial and commercial products, actually locating customers is usually a matter of making telephone calls. In developing such products within an existing firm, a field sales force can often provide names of customers, although the team must be careful about biasing the selection of customers toward those with allegiances to a particular manufacturer. The telephone book can be used to identify names of some types of customers for some classes of products (e.g., building contractors or insurance agents). For products that are integral to a customer's job, getting someone to agree to an interview is usually simple; these customers are anxious to discuss their needs. For consumer products, customers can also be located by making telephone calls. However, arranging a set of interviews for consumer products generally requires more inquiries than for industrial or commercial products because the benefit of participating in an interview is less direct for these customers.

The Art of Eliciting Customer Needs Data

The techniques we present here are aimed primarily at interviewing end users, but these methods do apply to all of the three data-gathering modes and to all types of stakeholders. The basic approach is to be receptive to information provided by customers and to avoid confrontations or defensive posturing. Gathering needs data is very different from a sales call: the goal is to elicit an honest expression of needs, not to convince a customer of what he or she needs. In most cases customer interactions will be verbal; interviewers ask questions and the customer responds. A prepared interview guide is valuable for structuring this dialogue. Some helpful questions and prompts for use after the interviewers introduce themselves and explain the purpose of the interview are:

- When and why do you use this type of product?
- Walk us through a typical session using the product.
- What do you like about the existing products?
- What do you dislike about the existing products?
- What issues do you consider when purchasing the product?
- What improvements would you make to the product?

Here are some general hints for effective interaction with customers:

- *Go with the flow.* If the customer is providing useful information, do not worry about conforming to the interview guide. The goal is to gather important data on customer needs, not to complete the interview guide in the allotted time.
- *Use visual stimuli and props.* Bring a collection of existing and competitors' products, or even products that are tangentially related to the product under development. At the end of a session, the interviewers might even show some preliminary product concepts to get customers' early reactions to various approaches.
- *Suppress preconceived hypotheses about the product technology.* Frequently customers will make assumptions about the product concept they expect would meet their needs. In these situations, the interviewers should avoid biasing the discussion with assumptions about how the product will eventually be designed or pro-

duced. When customers mention specific technologies or product features, the interviewer should probe for the underlying need the customer believes the suggested solution would satisfy.

- *Have the customer demonstrate the product and/or typical tasks related to the product.* If the interview is conducted in the use environment, a demonstration is usually convenient and invariably reveals new information.
- *Be alert for surprises and the expression of latent needs.* If a customer mentions something surprising, pursue the lead with follow-up questions. Frequently, an unexpected line of questioning will reveal *latent needs*—important dimensions of the customers' needs that are neither fulfilled nor commonly articulated and understood.
- *Watch for nonverbal information.* The process described in the chapter is aimed at developing better physical products. Unfortunately, words are not always the best way to communicate needs related to the physical world. This is particularly true of needs involving the human dimensions of the product, such as comfort, image, or style. The development team must be constantly aware of the nonverbal messages provided by customers. What are their facial expressions? How do they hold competitors' products?

Note that many of our suggested questions and guidelines assume that the customer has some familiarity with products similar to the new product under development. This is almost always true. For example, even before the first cordless screwdriver became available, people installed fasteners. Developing an understanding of customer needs as they relate to the general fastening task would still have been beneficial in developing the first cordless tool. Similarly, understanding the needs of customers using other types of cordless appliances, such as electric razors, would also have been useful. We can think of no product so revolutionary that there would be no analogous products or tasks from which the development team could learn. However, in gathering needs relating to truly revolutionary products with which customers have no experience, the interview questions should be focused on the task or situation in which the new product will be applied, rather than on the product itself.

Documenting Interactions with Customers

Four methods are commonly used for documenting interactions with customers:

1. *Audio recording:* Making an audio recording of the interview is very easy. Unfortunately, transcribing the recording into text is very time consuming, and it can be expensive to hire someone to do it. Also, audio recording has the disadvantage of being intimidating to some customers.

2. *Notes:* Handwritten notes are the most common method of documenting an interview. Designating one person as the primary notetaker allows the other person to concentrate on effective questioning. The notetaker should strive to capture some of the wording of every customer statement verbatim. These notes, if transcribed immediately after the interview, can be used to create a description of the interview that is very close to an actual transcript. This debriefing immediately after the interview also facilitates sharing of insights between the interviewers.

3. *Video recording:* Video recording is almost always used to document a focus group session. It is also very useful for documenting observations of the customer in the use environment and/or using existing products. The video recording is useful for bringing new team members "up to speed" and is also useful as raw material for presentations to upper management. Multiple viewings of video recordings of customers in action often facilitate the identification of latent customer needs. Video recording is also useful for capturing many aspects of the end user's environment.

4. *Still photography:* Taking photographs provides many of the benefits of video recording. The primary advantages of still photography are ease of display of the photos, excellent image quality, and readily available equipment. The primary disadvantage is the relative inability to record dynamic information.

The final result of the data-gathering phase of the process is a set of raw data, usually in the form of *customer statements* but frequently supplemented by video recordings or photographs. A data template implemented in a spreadsheet is useful for organizing these raw data. Figure 3.6 shows an example of a portion of such a template. We recommend that the template be filled in as soon as possible after the interaction with the customer and edited by the other development team members present during the interaction. The first column in the main body of the template indicates the question or prompt that elicited the customer data. The second column is a list of verbatim statements the customer made or an observation of a customer action (from a video recording or from direct observation). The third column contains the customer needs implied by the raw data. Some emphasis should be placed on investigating clues that may identify potential latent needs. Such clues may be in the form of humorous remarks, less serious suggestions, frustrations, nonverbal information, or observations and descriptions of the use environment. The symbol (!) is used in Figure 3.6 to flag potential latent needs. Techniques for interpreting the raw data in terms of customer needs are given in the next section.

The final task in step 1 is to write thank-you notes to the customers involved in the process. Invariably, the team will need to solicit further customer information, so developing and maintaining a good rapport with a set of users is important.

3.2 | STEP 2: INTERPRET RAW DATA IN TERMS OF CUSTOMER NEEDS

Customer needs are expressed as written statements and are the result of interpreting the need underlying the raw data gathered from the customers. Each statement or observation (as listed in the second column of the data template) may be translated into any number of customer needs. Griffin and Hauser found that multiple analysts may translate the same interview notes into different needs, so it is useful to have more than one team member conducting the translation process. Below we provide five guidelines for writing need statements. The first two

Customer:	Bill Esposito	Interviewer(s):	Jonathan and Lisa
Address:	100 Memorial Drive	Date:	19 December 1999
	Cambridge, MA 02139		
Telephone:	617-864-1274	Currently uses:	Craftsman Model A3
Willing to do follow-up?	Yes	Type of user:	Building maintenance

Question/Prompt	Customer Statement	Interpreted Need
Typical uses	I need to drive screws fast, faster than by hand.	The SD drives screws faster than by hand.
	I sometimes do duct work; use sheet metal screws.	The SD drives sheet metal screws into metal duct work.
	A lot of electrical; switch covers, outlets, fans, kitchen appliances.	The SD can be used for screws on electrical devices.
Likes—current tool	I like the pistol grip; it feels the best.	The SD is comfortable to grip.
	I like the magnetized tip.	The SD tip retains the screw before it is driven.
Dislikes—current tool	I don't like it when the tip slips off the screw.	The SD tip remains aligned with the screw head without slipping.
	I would like to be able to lock it so I can use it with a dead battery.	The user can apply torque manually to the SD to drive a screw. (!)
	Can't drive screws into hard wood.	The SD can drive screws into hard wood.
	Sometimes I strip tough screws.	The SD does not strip screw heads.
Suggested improvements	An attachment to allow me to reach down skinny holes.	The SD can access screws at the end of deep, narrow holes.
	A point so I can scrape paint off of screws.	The SD allows the user to work with screws that have been painted over.
	Would be nice if it could punch a pilot hole.	The SD can be used to create a pilot hole. (!)

Figure 3.6
Customer data template filled in with sample customer statements and interpreted needs. SD is an abbreviation for screwdriver. (Note that this template represents a partial list from a single interview. A typical interview session may elicit more than 50 customer statements and interpreted needs.)

Guideline	Customer Statement	Need Statement—Right	Need Statement—Wrong
"What" not "how"	"Why don't you put protective shields around the battery contacts?"	The screwdriver battery is protected from accidental shorting.	The screwdriver battery contacts are covered by a plastic sliding door.
Specificity	"I drop my screwdriver all the time."	The screwdriver operates normally after repeated dropping.	The screwdriver is rugged.
Positive not negative	"It doesn't matter if it's raining; I still need to work outside on Saturdays."	The screwdriver operates normally in the rain.	The screwdriver is not disabled by the rain.
An attribute of the product	"I'd like to charge my battery from my cigarette lighter."	The screwdriver battery can be charged from an automobile cigarette lighter.	An automobile cigarette lighter adapter can charge the screwdriver battery.
Avoid "must" and "should"	"I hate it when I don't know how much juice is left in the batteries of my cordless tools."	The screwdriver provides an indication of the energy level of the battery.	The screwdriver should provide an indication of the energy level of the battery.

Figure 3.7
Examples illustrating the guidelines for writing need statements.

guidelines are fundamental and are critical to effective translation; the remaining three guidelines ensure consistency of phrasing and style across all team members. Figure 3.7 provides examples to illustrate each guideline.

- *Express the need in terms of* **what** *the product has to do, not in terms of* **how** *it might do it.* Customers often express their preferences by describing a solution concept or an implementation approach; however, the need statement should be expressed in terms independent of a particular technological solution.
- *Express the need as specifically as the raw data.* Needs can be expressed at many different levels of detail. To avoid loss of information, express the need at the same level of detail as the raw data.
- *Use positive, not negative, phrasing.* Subsequent translation of a need into a product specification is easier if the need is expressed as a positive statement. This is not a rigid guideline, because sometimes positive phrasing is difficult and awkward. For example, one of the need statements in Figure 3.6 is "the screwdriver does not strip screw heads." This need is more naturally expressed in a negative form.
- *Express the need as an attribute of the product.* Wording needs as statements about the product ensures consistency and facilitates subsequent translation into product specifications. Not all needs can be cleanly expressed as attributes of the product, however, and in most of these cases the needs can be expressed as attributes of the user of the product (e.g., "the user can apply torque manually to the screwdriver to drive a screw").

- *Avoid the words* **must** *and* **should.** The words *must* and *should* imply a level of importance for the need. Rather than casually assigning a binary importance rating (*must* versus *should*) to the needs at this point, we recommend deferring the assessment of the importance of each need until step 4.

The list of customer needs is the superset of all the needs elicited from all the interviewed customers in the target market. Some needs may not be technologically realizable. The constraints of technical and economic feasibility are incorporated into the process of establishing product specifications in subsequent development steps. (See Chapter 5, Product Specifications.) In some cases customers will have expressed conflicting needs. At this point in the process the team does not attempt to resolve such conflicts, but simply documents both needs. Deciding how to address conflicting needs is one of the challenges of the subsequent concept development activities.

3.3 | STEP 3: ORGANIZE THE NEEDS INTO A HIERARCHY

The result of steps 1 and 2 should be a list of 50 to 300 *need statements*. Such a large number of detailed needs is awkward to work with and difficult to summarize for use in subsequent development activities. The goal of step 3 is to organize these needs into a hierarchical list. The list will typically consist of a set of *primary needs,* each one of which will be further characterized by a set of *secondary*

The SD provides plenty of power to drive screws.
*   The SD maintains power for several hours of heavy use.
**   The SD can drive screws into hardwood.
      The SD drives sheet metal screws into metal ductwork.
***   The SD drives screws faster than by hand.

The SD makes it easy to start a screw.
*   The SD retains the screw before it is driven.
*!   The SD can be used to create a pilot hole.

The SD works with a variety of screws.
**   The SD can turn phillips, torx, socket, and hex head screws.
**   The SD can turn many sizes of screws.

The SD can access most screws.
      The SD can be maneuvered in tight areas.
**   The SD can access screws at the end of deep, narrow holes.

The SD turns screws that are in poor condition.
      The SD can be used to remove grease and dirt from screws.
      The SD allows the user to work with painted screws.

The SD feels good in the user's hand.
***   The SD is comfortable when the user pushes on it.
***   The SD is comfortable when the user resists twisting.
*   The SD is balanced in the user's hand.
!   The SD is equally easy to use in right or left hands.
      The SD weight is just right.
      The SD is warm to touch in cold weather.
      The SD remains comfortable when left in the sun.

The SD is easy to control while turning screws.
***   The user can easily push on the SD.
***   The user can easily resist the SD twisting.
      The SD can be locked "on."
**!   The SD speed can be controlled by the user while turning a screw.
*   The SD remains aligned with the screw head without slipping.
**   The user can easily see where the screw is.
*   The SD does not strip screw heads.
*   The SD is easily reversible.

The SD is easy to set up and use.
*   The SD is easy to turn on.
*   The SD prevents inadvertent switching off.
*   The user can set the maximum torque of the SD.
*!   The SD provides ready access to bits or accessories.
*   The SD can be attached to the user for temporary storage.

The SD power is convenient.
*   The SD is easy to recharge.
      The SD can be used while recharging.
***   The SD recharges quickly.
      The SD batteries are ready to use when new.
**!   The user can apply torque manually to the SD to drive a screw.

The SD lasts a long time.
**   The SD tip survives heavy use.
      The SD can be hammered.
*   The SD can be dropped from a ladder without damage.

The SD is easy to store.
*   The SD fits in a toolbox easily.
**   The SD can be charged while in storage.
      The SD resists corrosion when left outside or in damp places.
*!   The SD maintains its charge after long periods of storage.
      The SD maintains its charge when wet.

The SD prevents damage to the work.
*   The SD prevents damage to the screw head.

      The SD prevents scratching of finished surfaces.

The SD has a pleasant sound when in use.

The SD looks like a professional quality tool.

The SD is safe.
      The SD can be used on electrical devices.
***   The SD does not cut the user's hands.

Figure 3.8
Hierarchical list of primary and secondary customer needs for the cordless screwdriver. Importance ratings for the secondary needs are indicated by the number of *'s, with *** denoting critically important needs. Latent needs are denoted by !.

needs. In cases of very complex products, the secondary needs may be broken down into tertiary needs as well. The primary needs are the most general needs, while the secondary and tertiary needs express needs in more detail. Figure 3.8 shows the resulting hierarchical list of needs for the screwdriver example. For the screwdriver, there are 15 primary needs and 49 secondary needs. Note that two of the primary needs have no associated secondary needs.

The procedure for organizing the needs into a hierarchical list is intuitive, and many teams can successfully complete the task without detailed instructions. For com-

pleteness, we provide a step-by-step procedure here. This activity is best performed on a large table by a small group of team members.

1. ***Print or write each need statement on a separate card or self-stick note.*** A print macro can be easily written to print the need statements directly from the data template. A nice feature of this approach is that the need can be printed in a large font in the center of the card and then the original customer statement and other relevant information can be

printed in a small font at the bottom of the card for easy reference. Four cards can be cut from a standard printed sheet.

2. ***Eliminate redundant statements.*** Those cards expressing redundant need statements can be stapled together and treated as a single card. Be careful to consolidate only those statements that are identical in meaning.

3. ***Group the cards according to the similarity of the needs they express.*** At this point, the team should attempt to create groups of roughly three to seven cards that express similar needs. The logic by which groups are created deserves special attention. Novice development teams often create groups according to a technological perspective, clustering needs relating to, for example, materials, packaging, or power. Or they create groups according to assumed physical components such as enclosure, bits, switch, and battery. Both of these approaches are dangerous. Recall that the goal of the process is to create a description of the needs of the customer. For this reason, the groupings should be consistent with the way customers think about their needs and not with the way the development team thinks about the product. The groups should correspond to needs customers would view as similar. In fact, some practitioners use a process in which customers actually organize the need statements.

4. ***For each group, choose a label.*** The label is itself a statement of need that generalizes all of the needs in the group. It can be selected from one of the needs in the group, or the team can write a new need statement.

5. ***Consider creating supergroups consisting of two to five groups.*** If there are fewer than 20 groups, then a two-level hierarchy is probably sufficient to organize the data. In this case, the group labels are primary needs and the group members are secondary needs. However, if there are more than 20 groups, the team may consider creating supergroups, and therefore a third level in the hierarchy. The process of creating supergroups is identical to the process of creating groups. As with the previous step, cluster groups according to similarity of the need they express and then create or select a supergroup label. These supergroup labels become the primary needs, the group labels become the secondary needs, and the members of the groups become tertiary needs.

6. ***Review and edit the organized needs statements.*** The arrangement of needs in a hierarchy is not unique in terms of being correct. At this point, the team may wish to consider alternative groupings or labels and may engage another group to suggest alternative arrangements.

The process is more complicated when the team attempts to reflect the needs of two or more distinct market segments. There are at least two approaches that can be taken to address this challenge. First, the team can label each need with the segment (and possibly the name) of the customer from whom the need was elicited. This way, differences in needs across segments can be observed directly. One practical, visual technique for this labeling is to use different colors of paper for the cards on which the needs statements are written, with each color corresponding to a different market segment. The other approach to multiple market segments is to perform the clustering process separately for each market segment. Using this approach, the team can observe differences both in the needs themselves and in the ways in which these needs are best organized. We recommend that the team adopt this parallel, independent approach when the segments are very different in their needs and when there is some doubt about the ability of the team to address the different segments with the same product.

3.4 | STEP 4: ESTABLISH THE RELATIVE IMPORTANCE OF THE NEEDS

The hierarchical list alone does not provide any information on the relative importance that customers place on different needs. Yet the development team will have to make trade-offs and allocate resources in designing the product. A sense of the relative importance of the various needs is essential to making these trade-offs correctly. Step 4 in the needs process establishes the relative importance of the customer needs identified in steps 1 through 3. The outcome of this step is a numerical importance weighting for a subset of the needs. There are two basic approaches to the task: (1) relying on the consensus of the team members based on their experience with customers, or (2) basing the importance assessment on further customer surveys. The obvious trade-off between the two approaches is cost and speed versus accuracy: the team can make an educated assessment of the relative importance of the needs in one meeting, while a customer survey generally takes a minimum of two weeks. In most cases we

---

Cordless Screwdriver Survey

For each of the following cordless screwdriver features, please indicate on a scale of 1 to 5 how important the feature is to you. Please use the following scale:

1. Feature is undesirable. I would not consider a product with this feature.
2. Feature is not important, but I would not mind having it.
3. Feature would be nice to have, but is not necessary.
4. Feature is highly desirable, but I would consider a product without it.
5. Feature is critical. I would not consider a product without this feature.

Also indicate by checking the box to the right if you feel that the feature is unique, exciting, and/or unexpected.

Importance of feature on scale of 1 to 5	Check box if feature is unique, exciting, and/or unexpected.
_____ The screwdriver maintains power for several hours of heavy use.	❏
_____ The screwdriver can drive screws into hardwood.	❏
_____ The screwdriver speed can be controlled by the user while turning a screw.	❏
_____ The screwdriver has a pleasant sound when in use.	❏

And so forth.

Figure 3.9
Example importance survey (partial).

believe the customer survey is important and worth the time required to complete it. Other development tasks, such as concept generation and analysis of competitive products, can begin before the relative importance surveys are complete.

The team should at this point have developed a rapport with a group of customers. These same customers can be surveyed to rate the relative importance of the needs that have been identified. The survey can be done in person, by telephone, or by mail. Few customers will respond to a survey asking them to evaluate the importance of 100 needs, so typically the team will work with only a subset of the needs. A practical limit on how many needs can be addressed in a customer survey is about 50. This limit is not too severe, however, because many of the needs are either obviously important (e.g., the screwdriver fits in a toolbox easily) or are easy to implement (e.g., the screwdriver prevents inadvertent switching off). The team can therefore limit the scope of the survey by querying customers only about needs that are likely to give rise to difficult technical trade-offs or costly features in the product design. Such needs would include the need to vary speed, the need to drive screws into hardwood, and the need to have the screwdriver emit a pleasant sound. Alternatively the team could develop a set of surveys to ask a variety of customers each about different subsets of the needs list. There are many survey designs for establishing the relative importance of customer needs. One good design is illustrated by the portion of the cordless screwdriver survey shown in Figure 3.9. In addition to asking for importance ratings, this survey asks the respondent to explicitly identify the needs that are unique or unexpected. This information can be used to help the team identify latent needs.

The survey responses for each need statement can be characterized in a variety of ways: by the mean, by the standard deviation, or by the number of responses in each category. The responses can then be used to assign an importance weighting to the need statements. The same scale of 1 to 5 can be used to summarize the importance data. The needs in Figure 3.8 are rated according to the survey data, with the importance ratings denoted by the number of *'s next to each need statement and the latent needs denoted by !. Note that no critical needs are also latent needs. This is because if a need were critical, customers would not be surprised or excited by it; they would expect it to be met.

3.5 | STEP 5: REFLECT ON THE RESULTS AND THE PROCESS

The final step in the method is to reflect on the results and the process. While the process of identifying customer needs can be usefully structured, it is not an exact science. The team must challenge its results to verify that they are consistent with the knowledge and intuition the team has developed through many hours of interaction with customers. Some questions to ask include:

- Have we interacted with all of the important types of customers in our target market?
- Are we able to see beyond needs related only to existing products in order to capture the latent needs of our target customers?
- Are there areas of inquiry we should pursue in follow-up interviews or surveys?
- Which of the customers we spoke to would be good participants in our on-going development efforts?
- What do we know now that we didn't know when we started? Are we surprised by any of the needs?
- Did we involve everyone within our own organization who needs to deeply understand customer needs?
- How might we improve the process in future efforts?

- Customer needs should be expressed in terms of what the product has to do, not in terms of how the product might be implemented. Adherence to this principle leaves the development team with maximum flexibility to generate and select product concepts.
- The key benefits of the method are: ensuring that the product is focused on customer needs and that no critical customer need is forgotten; developing a clear understanding among members of the development team of the needs of the customers in the target market; developing a fact base to be used in generating concepts, selecting a product concept, and establishing product specifications; and creating an archival record of the needs phase of the development process.

3.6 | SUMMARY

Identifying customer needs is an integral part of the concept development phase of the product development process. The resulting customer needs are used to guide the team in establishing product specifications, generating product concepts, and selecting a product concept for further development.

- The process of identifying customer needs includes five steps:
  1. Gather raw data from customers.
  2. Interpret the raw data in terms of customer needs.
  3. Organize the needs into a hierarchy of primary and secondary needs.
  4. Establish the relative importance of the needs.
  5. Reflect on the results and the process.
- Creating a high-quality information channel from customers to the product developers ensures that those who directly control the details of the product, including the product designers, fully understand the needs of the customer.
- Lead users are a good source of customer needs because they experience new needs months or years ahead of most customers and because they stand to benefit substantially from new product innovations. Furthermore, they are frequently able to articulate their needs more clearly than typical customers.
- Latent needs may be even more important than explicit needs in determining customer satisfaction. Latent needs are those that many customers recognize as important in a final product but do not or cannot articulate in advance.

3.7 | REFERENCES AND BIBLIOGRAPHY

Many current resources are available on the Internet via

www.ulrich-eppinger.net

Concept engineering is a method developed by Burchill at MIT in collaboration with the Center for Quality of Management. This chapter benefits from our observations of the development and application of concept engineering. For a complete and detailed description of concept engineering, see:

Burchill, Gary, et al., *Concept Engineering,* Center for Quality of Management, Cambridge, MA, Document No. ML0080, 1997.

The research by Griffin and Hauser is one of the only rigorous efforts to validate different methods for extracting needs from interview data. Their study of the fraction of needs identified as a function of the number of customers interviewed is particularly interesting.

Griffin, Abbie, and John R. Hauser, "The Voice of the Customer," *Marketing Science,* Vol. 12, No. 1, Winter 1993, pp.1-27.

Kinnear and Taylor thoroughly discuss data collection methods and survey design.

Kinnear, Thomas C., and James R. Taylor, *Marketing Research: An Applied Approach,* fifth edition, McGraw-Hill, New York, NY, 1995.

Norman has written extensively on user needs, especially as related to the cognitive challenges of using products.

Norman, Donald A., *The Design of Everyday Things,* Doubleday, New York, 1990.

Payne's book is a detailed and interesting discussion of how to pose questions in surveys.

Payne, Stanley L., *The Art of Asking Questions,* Princeton University Press, Princeton, NJ, 1980.

Total quality management (TQM) provides a valuable perspective on how identifying customer needs fits into an overall effort to improve the quality of goods and services.

Shiba, Shoji, Alan Graham, and David Walden, *A New American TQM: Four Practical Revolutions in Management,* Productivity Press, Cambridge, MA, and The Center for Quality of Management, Cambridge, MA, 1993.

Urban and Hauser provide a thorough discussion of how to create hierarchies of needs (along with many other topics).

Urban, Glen L., and John R. Hauser, *Design and Marketing of New Products,* second edition, Prentice Hall, Englewood Cliffs, NJ, 1993.

Von Hippel describes many years of research on the role of lead users in innovation. He provides useful guidelines for identifying lead users.

von Hippel, Eric, *The Sources of Innovation,* Oxford University Press, New York, NY, 1988.

Exercises

1. Translate the following customer statements about a student book bag into proper needs statements:
   a. "See how the leather on the bottom of the bag is all scratched; it's ugly."
   b. "When I'm standing in line at the cashier trying to find my checkbook while balancing my bag on my knee, I feel like a stork."
   c. "This bag is my life; if I lose it I'm in big trouble."
   d. "There's nothing worse than a banana that's been squished by the edge of a textbook."
   e. "I never use both straps on my knapsack; I just sling it over one shoulder."

2. Observe someone performing an everyday task. (Ideally, you should choose a task for which you can observe different users performing the task repeatedly.) Identify frustrations and difficulties encountered by these people. Identify the latent customer needs.

3. Choose a product that continually annoys you. Identify the needs the developers of this product missed. Why do you think these needs were not met? Do you think the developers deliberately ignored these needs?

Thought Questions

1. One of the reasons the method is effective is that it involves the entire development team. Unfortunately, the method can become unwieldy with a team of more than 10 people. How might you modify the method to maximize involvement yet maintain a focused and decisive effort given a large development team?

2. Can the process of identifying customer needs lead to the creation of innovative product concepts? In what ways? Could a structured process of identifying customer needs lead to a fundamentally new product concept like the Post-It note?

Product Specifications

4

Specialized Bicycle Components was interested in developing a front suspension fork for the mountain bike market. Although the firm was already selling a suspension fork (Figure 4.1), it was successful primarily in the high-performance segment of the market—racing cyclists with less concern for cost or long-term durability. The firm wished to broaden the sales of suspension forks and therefore was interested in developing a product that would provide high value for the recreational cyclist.

The development team had spent a great deal of time identifying customer needs. In addition to logging many hours of riding on suspended bikes themselves, the members of the team had interviewed lead users at mountain bike races and recreational cyclists on local trails, and they

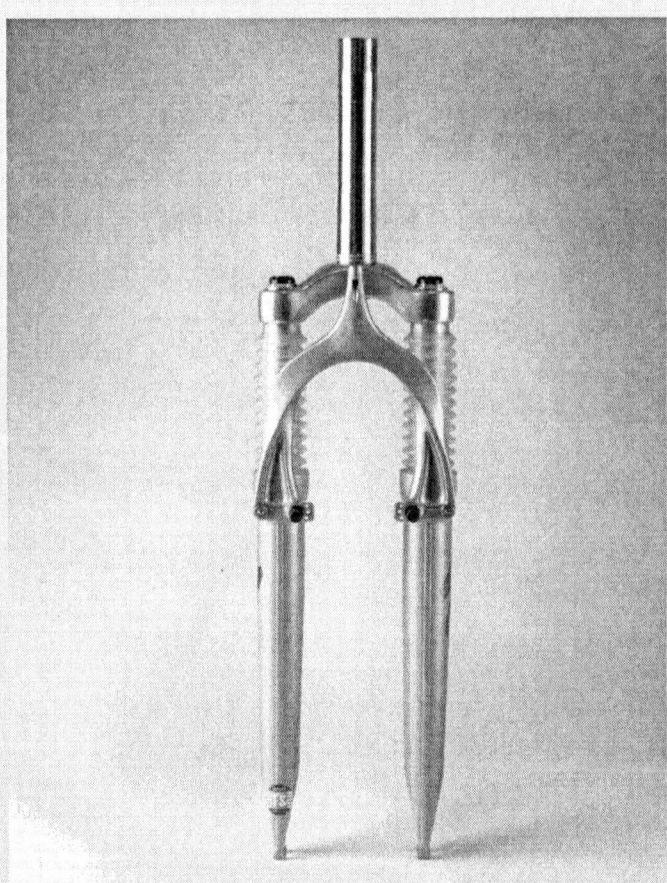

Figure 4.1
One of Specialized's existing suspension forks. Stuart Cohen

also had spent time working with dealers in their stores. As a result of this process they had assembled a list of customer needs. They now faced several challenges:

- How could the relatively subjective customer needs be translated into precise targets for the remaining development effort?
- How could the team and its senior management agree on what would constitute success or failure of the resulting product design?
- How could the team develop confidence that its intended product would garner a substantial share of the suspension fork market?
- How could the team resolve the inevitable trade-offs among product characteristics like cost and weight?

This chapter presents a method for establishing product specifications. We assume that the customer needs are already documented as described in Chapter 4, Identifying Customer Needs. The method employs several simple information systems, all of which can be constructed using conventional spreadsheet software.

4.1 | WHAT ARE SPECIFICATIONS?

Customer needs are generally expressed in the "language of the customer." The primary customer needs for the suspension fork are listed in Figure 4.2. Customer needs such as "the suspension is easy to install" or "the suspension enables high-speed descents on bumpy trails" are typical in terms of the subjective quality of the expressions. However, while such expressions are helpful in developing a clear sense of the issues of interest to customers, they provide little specific guidance about how to design and engineer the product. They simply leave too much margin for subjective interpretation. For this reason, development teams usually establish a set of specifications, which spell out in precise, measurable detail *what* the product has to do. Product specifications do not tell the team *how* to address the customer needs, but they do represent an unambiguous agreement on what the team will attempt to achieve in order to satisfy the customer needs. For example, in contrast to the customer need that "the suspension is easy to install," the corresponding specification might

No.		Need	Imp.
1	The suspension	reduces vibration to the hands.	3
2	The suspension	allows easy traversal of slow, difficult terrain.	2
3	The suspension	enables high-speed descents on bumpy trails.	5
4	The suspension	allows sensitivity adjustment.	3
5	The suspension	preserves the steering characteristics of the bike.	4
6	The suspension	remains rigid during hard cornering.	4
7	The suspension	is lightweight.	4
8	The suspension	provides stiff mounting points for the brakes.	2
9	The suspension	fits a wide variety of bikes, wheels, and tires.	5
10	The suspension	is easy to install.	1
11	The suspension	works with fenders.	1
12	The suspension	instills pride.	5
13	The suspension	is affordable for an amateur enthusiast.	5
14	The suspension	is not contaminated by water.	5
15	The suspension	is not contaminated by grunge.	5
16	The suspension	can be easily accessed for maintenance.	3
17	The suspension	allows easy replacement of worn parts.	1
18	The suspension	can be maintained with readily available tools.	3
19	The suspension	lasts a long time.	5
20	The suspension	is safe in a crash.	5

Figure 4.2
Customer needs for the suspension fork and their relative importance (shown in a convenient spreadsheet format).

be that "the average time to assemble the fork to the frame is less than 75 seconds."

We intend the term *product specifications* to mean the precise description of what the product has to do. Some firms use the terms "product requirements" or "engineering characteristics" in this way. Other firms use "specifications" or "technical specifications" to refer to key design variables of the product such as the oil viscosity or spring constant of the suspension system. These are just differences in terminology. For clarity, let us be precise about a few definitions. A *specification* (singular) consists of a *metric* and a *value.* For example, "average time to assemble" is a metric, while "less than 75 seconds" is the value of this metric. Note that the value may take on several forms, including a particular number, a range, or an inequality. Values are always labeled with the appropriate units (e.g., seconds, kilograms, joules). Together, the metric and value form a specification. The *product specifications* (plural) are simply the set of the individual specifications.

4.2 | WHEN ARE SPECIFICATIONS ESTABLISHED?

In an ideal world, the team would establish the product specifications once early in the development process and then proceed to design and engineer the product to exactly meet those specifications. For some products, such as soap or soup, this approach works quite well; the technologists on the team can reliably concoct a formulation that satisfies almost any reasonable specifications. However, for technology-intensive products this is rarely possible. For such products, specifications are established at least twice. Immediately after identifying the customer needs, the team sets *target specifications.* These specifications represent the hopes and aspirations of the team, but they are established before the team knows what constraints the

product technology will place on what can be achieved. The team's efforts may fail to meet some of these specifications and may exceed others, depending on the product concept the team eventually selects. For this reason, the target specifications must be refined after a product concept has been selected. The team revisits the specifications while assessing the actual technological constraints and the expected production costs. To set the *final specifications,* the team must frequently make hard trade-offs among different desirable characteristics of the product. For simplicity, we present a two-stage process for establishing specifications, but we note that in some organizations specifications are revisited many times throughout the development process.

The two stages in which specifications are established are shown as part of the concept development process in Figure 4.3. Note that the final specifications are one of the key elements of the development plan, which is usually documented in the project's *contract book.* The contract book (described in Chapter 14, Managing Projects) specifies what the team agrees to achieve, the project schedule, the required resources, and the economic implications for the business. The list of product specifications is also one of the key information systems used by the team throughout the development process.

This chapter presents two methods: the first is for establishing the target specifications and the second is for setting the final specifications after the product concept has been selected.

4.3 | ESTABLISHING TARGET SPECIFICATIONS

As Figure 4.3 illustrates, the target specifications are established after the customer needs have been identified but before product concepts have been generated and the most

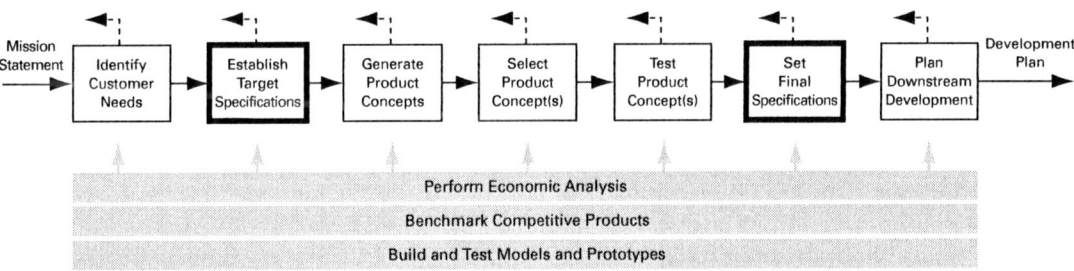

Figure 4.3
The concept development process. The target specifications are set early in the process, but setting the final specifications must wait until after the product concept has been selected.

promising one(s) selected. An arbitrary setting of the specifications may not be technically feasible. For example, in designing a suspension fork, the team cannot assume in advance that it will be able to achieve simultaneously a mass of 1 kilogram, a manufacturing cost of $30, and the best descent time on the test track, as these are three quite aggressive specifications. Actually meeting the specifications established at this point is contingent upon the details of the product concept the team eventually selects. For this reason, such preliminary specifications are labeled "target specifications." They are the goals of the development team, describing a product that the team believes would succeed in the marketplace. Later these specifications will be refined based on the limitations of the product concept actually selected.

The process of establishing the target specifications contains four steps:

1. Prepare the list of metrics, using the needs-metrics matrix, if necessary.
2. Collect the competitive benchmarking information.
3. Set ideal and marginally acceptable target values for each metric.
4. Reflect on the results and the process.

Step 1: Prepare the List of Metrics

The most useful metrics are those that reflect as directly as possible the degree to which the product satisfies the customer needs. The relationship between needs and metrics is central to the entire concept of specifications. The working assumption is that a translation from customer needs to a set of precise, measurable specifications is possible and that meeting specifications will therefore lead to satisfaction of the associated customer needs.

A list of metrics is shown in Figure 4.4. A good way to generate the list of metrics is to contemplate each need in turn and to consider what precise, measurable characteristic of the product will reflect the degree to which the product satisfies that need. In the ideal case, there is one and only one metric for each need. In practice, this is frequently not possible.

For example, consider the need that the suspension be "easy to install." The team may conclude that this need is largely captured by measuring the time required for assembly of the fork to the frame. However, note the possible subtleties in this translation. Is assembly time really identical to ease of installation? The installation could be extremely fast but require an awkward and painful set of finger actions, which ultimately may lead to worker injury

or dealer frustration. Because of the imprecise nature of the translation process, those establishing the specifications should have been directly involved in identifying the customer needs. In this way the team can rely on its understanding of the meaning of each need statement derived from firsthand interactions with customers.

The need for the fork to reduce vibration to the user's hands may be even more difficult to translate into a single metric, because there are many different conditions under which vibration can be transmitted, including small bumps on level roads and big bumps on rough trails. The team may conclude that several metrics are required to capture this need, including, for example, the metrics "attenuation from dropout to handlebar at 10 Hz" and "maximum value from the Monster." (The "Monster" is a shock test developed by *Mountain Bike* magazine.)

A simple needs-metrics matrix represents the relationship between needs and metrics. An example needs-metrics matrix is shown in Figure 4.5. The rows of the matrix correspond to the customer needs, and the columns of the matrix correspond to the metrics. A mark in a cell of the matrix means that the need and the metric associated with the cell are related; performance relative to the metric will influence the degree to which the product satisfies the customer need. This matrix is a key element of the *House of Quality,* a graphical technique used in *Quality Function Deployment,* or *QFD* (Hauser and Clausing, 1988). In many cases, we find the information in the needs-metrics matrix is just as easily communicated by listing the numbers of the needs related to each metric alongside the list of metrics (the second column in Figure 4.4). There are some cases, however, in which the mapping from needs to metrics is complex, and the matrix can be quite useful for representing this mapping.

A few guidelines should be considered when constructing the list of metrics:

- *Metrics should be complete.* Ideally each customer need would correspond to a single metric, and the value of that metric would correlate perfectly with satisfaction of that need. In practice, several metrics may be necessary to completely reflect a single customer need.
- *Metrics should be dependent, not independent, variables.* This guideline is a variant of the *what-not-how* principle introduced in Chapter 4. As do customer needs, specifications also indicate *what* the product must do, but not *how* the specifications will be achieved. Designers use many types of variables in product development; some are *dependent,* such as the mass of the fork, and some are *independent,* such as the material used for the fork. In other words, designers cannot control mass directly because it arises from other indepen-

dent decisions the designers will make, such as dimensions and materials choices. Metrics specify the overall performance of a product and should therefore be the dependent variables (i.e., the performance measures or output variables) in the design problem. By using dependent variables for the specifications, designers are left with the freedom to achieve the specifications using the best approach possible.

- *Metrics should be practical.* It does not serve the team to devise a metric for a bicycle suspension that can only be measured by a scientific laboratory at a cost of $100,000. Ideally, metrics will be directly observable or analyzable properties of the product that can be easily evaluated by the team.

- *Some needs cannot easily be translated into quantifiable metrics.* The need that the suspension instills pride may be quite critical to success in the fashion-conscious mountain bike market, but how can pride be quantified? In these cases, the team simply repeats the need statement as a specification and notes that the metric is subjective and would be evaluated by a panel of customers. (We indicate this by entering "Subj." in the units column.)

- *The metrics should include the popular criteria for comparison in the marketplace.* Many customers in various markets buy products based on independently published evaluations. Such evaluations are found in, for example, *Popular Science, Consumer Reports,* or, in our case, *Bicycling* and *Mountain Bike* magazines. If

Metric No.	Need Nos.	Metric	Imp.	Units
1	1, 3	Attenuation from dropout to handlebar at 10 Hz	3	dB
2	2, 6	Spring preload	3	N
3	1, 3	Maximum value from the Monster	5	g
4	1, 3	Minimum descent time on test track	5	s
5	4	Damping coefficient adjustment range	3	N-s/m
6	5	Maximum travel (26-in. wheel)	3	mm
7	5	Rake offset	3	mm
8	6	Lateral stiffness at the tip	3	kN/m
9	7	Total mass	4	kg
10	8	Lateral stiffness at brake pivots	2	kN/m
11	9	Headset sizes	5	in.
12	9	Steertube length	5	mm
13	9	Wheel sizes	5	List
14	9	Maximum tire width	5	in.
15	10	Time to assemble to frame	1	s
16	11	Fender compatibility	1	List
17	12	Instills pride	5	Subj.
18	13	Unit manufacturing cost	5	US$
19	14	Time in spray chamber without water entry	5	s
20	15	Cycles in mud chamber without contamination	5	k-cycles
21	16, 17	Time to disassemble/assemble for maintenance	3	s
22	17, 18	Special tools required for maintenance	3	List
23	19	UV test duration to degrade rubber parts	5	Hours
24	19	Monster cycles to failure	5	Cycles
25	20	Japan Industrial Standards test	5	Binary
26	20	Bending strength (frontal loading)	5	

Figure 4.4
List of metrics for the suspension. The relative importance of each metric and the units for the metric are also shown. "Subj." is an abbreviation indicating that a metric is subjective.

the team knows that its product will be evaluated by the trade media and knows what the evaluation criteria will be, then it should include metrics corresponding to these criteria. *Mountain Bike* magazine uses a test machine called the Monster, which measures the vertical acceleration (in g's) of the handlebars as a bicycle equipped with the fork runs over a block 50 millimeters tall. For this reason, the team included "maximum value from the Monster" as a metric. If the team cannot find a relationship between the criteria used by the media and the customer needs it has identified, then it should ensure that a need has not been overlooked and/or should work with the media to revise the crite-ria. In a few cases, the team may conclude that high performance in the media evaluations is in itself a customer need and choose to include a metric used by the media that has little intrinsic technical merit.

In addition to denoting the needs related to each metric, Figure 4.4 contains the units of measurement and an importance rating for each metric. The units of measurement are most commonly conventional engineering units such as kilograms and seconds. However, some metrics will not lend themselves to numerical values. The need that the suspension "works with fenders" is best translated into a specification listing the models of fenders with which the

Need	1 Attenuation from dropout to handlebar at 10 Hz	2 Spring preload	3 Maximum value from the Monster	4 Minimum descent time on test track	5 Damping coefficient adjustment range	6 Maximum travel (26-in. wheel)	7 Rake offset	8 Lateral stiffness at the tip	9 Total mass	10 Lateral stiffness at brake pivots	11 Headset sizes	12 Steertube length	13 Wheel sizes	14 Maximum tire width	15 Time to assemble to frame	16 Fender compatibility	17 Instills pride	18 Unit manufacturing cost	19 Time in spray chamber without water entry	20 Cycles in mud chamber without contamination	21 Time to disassemble/assemble for maintenance	22 Special tools required for maintenance	23 UV test duration to degrade rubber parts	24 Monster cycles to failure	25 Japan Industrial Standards test	26 Bending strength (frontal loading)
1 Reduces vibration to the hands	●		●	●																						
2 Allows easy traversal of slow, difficult terrain		●																								
3 Enables high-speed descents on bumpy trails	●		●	●																						
4 Allows sensitivity adjustment					●																					
5 Preserves the steering characteristics of the bike						●	●																			
6 Remains rigid during hard cornering	●							●																		
7 Is lightweight									●																	
8 Provides stiff mounting points for the brakes										●																
9 Fits a wide variety of bikes, wheels, and tires											●	●	●	●												
10 Is easy to install															●											
11 Works with fenders																●										
12 Instills pride																	●									
13 Is affordable for an amateur enthusiast																		●								
14 Is not contaminated by water																			●							
15 Is not contaminated by grunge																				●						
16 Can be easily accessed for maintenance																					●					
17 Allows easy replacement of worn parts																					●	●				
18 Can be maintained with readily available tools																						●				
19 Lasts a long time																							●	●		
20 Is safe in a crash																									●	●

Figure 4.5
The needs-metrics matrix.

fork is compatible. In this case, the value of the metric is actually a list of fenders rather than a number. For the metric involving the standard safety test, the value is pass/fail. (We indicate these two cases by entering "List" and "Binary" in the units column.)

The importance rating of a metric is derived from the importance ratings of the needs it reflects. For cases in which a metric maps directly to a single need, the importance rating of the need becomes the importance rating of the metric. For cases in which a metric is related to more than one need, the importance of the metric is determined by considering the importances of the needs to which it relates and the nature of these relationships. We believe that there are enough subtleties in this process that importance weightings can best be determined through discussion among the team members, rather than through a formal algorithm. When there are relatively few specifications and establishing the relative importance of these specifications is critically important, *conjoint analysis* may be useful. Conjoint analysis is described briefly later in this chapter and publications explaining the technique are referenced at the end of the chapter.

Step 2: Collect the Competitive Benchmarking Information

Unless the team expects to enjoy a total monopoly, the relationship of the new product to competitive products is paramount in determining commercial success. While the team will have entered the product development process with some idea of how it wishes to compete in the marketplace, the target specifications are the language the team uses to discuss and agree on the detailed positioning of its product relative to existing products, both its own and competitors'. Information on competing products must be gathered to support these positioning decisions.

An example of a competitive benchmarking chart is shown in Figure 4.6. The columns of the chart correspond to the competitive products and the rows are the metrics established in step 1. Note that the competitive benchmarking chart can be constructed as a simple appendage to the spreadsheet containing the list of metrics. (This information is one of the "rooms" in the *House of Quality,* described by Hauser and Clausing.)

The benchmarking chart is conceptually very simple. For each competitive product, the values of the metrics are simply entered down a column. Gathering these data can be very time consuming, involving (at the least) purchasing, testing, disassembling, and estimating the production costs of the most important competitive products. However, this investment of time is essential, as no product development team can expect to succeed without having this type of in-

formation. A word of warning: Sometimes the data contained in competitors' catalogs and supporting literature are not accurate. Where possible, values of the key metrics should be verified by independent testing or observation.

An alternative competitive benchmarking chart can be constructed with rows corresponding to the customer needs and columns corresponding to the competitive products (see Figure 4.7). This chart is used to compare customers' perceptions of the relative degree to which the products satisfy their needs. Constructing this chart requires collecting customer perception data, which can also be very expensive and time consuming. Some techniques for measuring customers' perceptions of satisfaction of needs are contained in a book by Urban and Hauser (1993). Both charts can be useful and any discrepancies between the two are instructive. At a minimum, a chart showing the competitive values of the metrics (Figure 4.6) should be created.

Step 3: Set Ideal and Marginally Acceptable Target Values for Each Metric

In this step, the team synthesizes the available information in order to actually set the *target values* for the metrics. Two types of target values are useful: an *ideal value* and a *marginally acceptable value.* The ideal value is the best result the team could hope for. The marginally acceptable value is the value of the metric that would just barely make the product commercially viable. Both of these targets are useful in guiding the subsequent stages of concept generation and concept selection, and for refining the specifications after the product concept has been selected.

There are five ways to express the values of the metrics:

- *At least X:* These specifications establish targets for the lower bound on a metric, but higher is still better. For example, the value of the brake mounting stiffness is specified to be at least 325 kilonewtons/meter.
- *At most X:* These specifications establish targets for the upper bound on a metric, with smaller values being better. For example, the value for the mass of the suspension fork is set to be at most 1.4 kilograms.
- *Between X and Y:* These specifications establish both upper and lower bounds for the value of a metric. For example, the value for the spring preload is set to be between 480 and 800 newtons. Any more and the suspension is harsh; any less and the suspension is too bouncy.
- *Exactly X:* These specifications establish a target of a particular value of a metric, with any deviation degrading performance. For example, the ideal value for the rake offset metric is set to 38 millimeters. This type of

Metric No.	Need Nos.	Metric	Imp.	Units	ST Tritrack	Maniray 2	Rox Tahx Quadra	Rox Tahx Ti 21	Tonka Pro	Gunhill Head Shox
1	1, 3	Attenuation from dropout to handlebar at 10Hz	3	dB	8	15	10	15	9	13
2	2, 6	Spring preload	3	N	550	760	500	710	480	680
3	1, 3	Maximum value from the Monster	5	g	3.6	3.2	3.7	3.3	3.7	3.4
4	1, 3	Minimum descent time on test track	5	s	13	11.3	12.6	11.2	13.2	11
5	4	Damping coefficient adjustment range	3	N-s/m	0	0	0	200	0	0
6	5	Maximum travel (26 in. wheel)	3	mm	28	48	43	46	33	38
7	5	Rake offset	3	mm	41.5	39	38	38	43.2	39
8	6	Lateral stiffness at the tip	3	kN/m	59	110	85	85	65	130
9	7	Total mass	4	kg	1.409	1.385	1.409	1.364	1.222	1.100
10	8	Lateral stiffness at brake pivots	2	kN/m	295	550	425	425	325	650
11	9	Headset sizes	5	in.	1.000 1.125	1.000 1.125 1.250	1.000 1.125	1.000 1.125 1.250	1.000 1.125	NA
12	9	Steertube length	5	mm	150 180 210 230 255	140 165 190 215	150 170 190 210	150 170 190 210 230	150 190 210 220	NA
13	9	Wheel sizes	5	List	26 in	26 in	26 in	26 in 700C	26 in	26 in
14	9	Maximum tire width	5	in.	1.5	1.75	1.5	1.75	1.5	1.5
15	10	Time to assemble to frame	1	s	35	35	45	45	35	85
16	11	Fender compatibility	1	List	Zefal	None	None	None	None	All
17	12	Instills pride	5	Subj.	1	4	3	5	3	5
18	13	Unit manufacturing cost	5	US$	65	105	85	115	80	100
19	14	Time in spray chamber without water entry	5	s	1300	2900	>3600	>3600	2300	>3600
20	15	Cycles in mud chamber without contamination	5	k-cycles	15	19	15	25	18	35
21	16, 17	Time to disassemble/ assemble for maintenance	3	s	160	245	215	245	200	425
22	17, 18	Special tools required for maintenance	3	List	Hex	Hex	Hex	Hex	Long hex	Hex, pin wrench
23	19	UV test duration to degrade rubber parts	5	Hours	400+	250	400+	400+	400+	250
24	19	Monster cycles to failure	5	Cycles	500k+	500k+	500k+	480k	500k+	330k
25	20	Japan Industrial Standards test	5	Binary	Pass	Pass	Pass	Pass	Pass	Pass
26	20	Bending strength (frontal loading)	5	kN	5.5	8.9	7.5	7.5	6.2	10.2

Figure 4.6
Competitive benchmarking chart based on metrics.

No.	Need	Imp.	ST Tritrack	Maniray 2	Rox Tahx Quadra	Rox Tahx Ti 21	Tonka Pro	Gunhill Head Shox
1	Reduces vibration to the hands	3	•	••••	••	•••••	••	•••
2	Allows easy traversal of slow, difficult terrain	2	••	••••	•••	•••••	•••	•••••
3	Enables high-speed descents on bumpy trails	5	•	•••••	••	•••••	••	•••
4	Allows sensitivity adjustment	3	•	••••	••	•••••	••	•••
5	Preserves the steering characteristics of the bike	4	••••	••	•	••	••••••••	
6	Remains rigid during hard cornering	4	•	•••	•	•••••	•	•••••
7	Is lightweight	4	•	•••	•	•••	••••	•••••
8	Provides stiff mounting points for the brakes	2	•	••••	•••	•••	•••••••	
9	Fits a wide variety of bikes, wheels, and tires	5	••••	•••••	•••	•••••	•••	•
10	Is easy to install	1	••••	•••••	••••	••••	•••••	•
11	Works with fenders	1	•••	•	•	•	•	•••••
12	Instills pride	5	•	••••	•••	•••••	•••	•••••
13	Is affordable for an amateur enthusiast	5	•••••	•	•••	•	•••	••
14	Is not contaminated by water	5	•	•••	••••	••••	••	•••••
15	Is not contaminated by grunge	5	•	•••	•	••••	••	•••••
16	Can be easily accessed for maintenance	3	••••	•••••	••••	••••	•••••	•
17	Allows easy replacement of worn parts	1	••••	•••••	••••	••••	•••••	•
18	Can be maintained with readily available tools	3	•••••	•••••	•••••	•••••	••	•
19	Lasts a long time	5	•••••	•••••	•••••	•••	•••••	•
20	Is safe in a crash	5	•••••	•••••	•••••	•••••	•••••	•••••

Figure 4.7
Competitive benchmarking chart based on perceived satisfaction of needs. (Scoring more "dots" corresponds to greater perceived satisfaction of the need.)

specification is to be avoided if possible because such specifications substantially constrain the design. Often, upon reconsideration, the team realizes that what initially appears as an "exactly X" specification can be expressed as a "between X and Y" specification.

- *A set of discrete values:* Some metrics will have values corresponding to several discrete choices. For example, the headset diameters are 1.000, 1.125, or 1.250 inches. (Industry practice is to use English units for these and several other critical bicycle dimensions.)

The desirable range of values for one metric may depend on another. In other words, we may wish to express a target as, for example, "the fork tip lateral stiffness is no more than 20 percent of the lateral stiffness at the brake pivots." In applications where the team feels this level of

complexity is warranted, such targets can easily be included, although we recommend that this level of complexity not be introduced until the final phase of the specifications process.

Using these five different types of expressions for values of the metrics, the team sets the target specifications. The team simply proceeds down the list of metrics and assigns both the marginally acceptable and ideal target values for each metric. These decisions are facilitated by the metric-based competitive benchmarking chart shown in Figure 4.6. To set the target values, the team has many considerations, including the capability of competing products available at the time, competitors' future product capabilities (if these are predictable), and the product's mission statement and target market segment. Figure 4.8 shows the targets assigned for the suspension fork.

Because most of the values are expressed in terms of bounds (upper or lower or both), the team is establishing the boundaries of the competitively viable product space. The team hopes that the product will meet some of the ideal targets but is confident that a product can be commercially viable even if it exhibits one or more marginally acceptable characteristics. Note that these specifications are preliminary because until a product concept is chosen and some of the design details are worked out, many of the exact trade-offs are quite uncertain.

Metric No.	Need Nos.	Metric	Imp.	Units	Marginal Value	Ideal Value
1	1, 3	Attenuation from dropout to handlebar at 10 Hz	3	dB	>10	>15
2	2, 6	Spring preload	3	N	480-800	650-700
3	1, 3	Maximum value from the Monster	5	g	<3.5	<3.2
4	1, 3	Minimum descent time on test track	5	s	<13.0	<11.0
5	4	Damping coefficient adjustment range	3	N-s/m	0	>200
6	5	Maximum travel (26-in. wheel)	3	mm	33-50	45
7	5	Rake offset	3	mm	37-45	38
8	6	Lateral stiffness at the tip	3	kN/m	>65	>130
9	7	Total mass	4	kg	<1.4	<1.1
10	8	Lateral stiffness at brake pivots	2	kN/m	>325	>650
11	9	Headset sizes	5	in.	1.000 1.125	1.000 1.125 1.250
12	9	Steertube length	5	mm	150 170 190 210	150 170 190 210 230
13	9	Wheel sizes	5	List	26 in.	26 in. 700C
14	9	Maximum tire width	5	in.	>1.5	>1.75
15	10	Time to assemble to frame	1	s	<60	<35
16	11	Fender compatibility	1	List	None	All
17	12	Instills pride	5	Subj.	>3	>5
18	13	Unit manufacturing cost	5	US$	<85	<65
19	14	Time in spray chamber without water entry	5	s	>2300	>3600
20	15	Cycles in mud chamber without contamination	5	k-cycles	>15	>35
21	16, 17	Time to disassemble/assemble for maintenance	3	s	<300	<160
22	17, 18	Special tools required for maintenance	3	List	Hex	Hex
23	19	UV test duration to degrade rubber parts	5	Hours	>250	>450
24	19	Monster cycles to failure	5	Cycles	>300k	>500k
25	20	Japan Industrial Standards test	5	Binary	Pass	Pass
26	20	Bending strength (frontal loading)	5	kN	>7.0	>10.0

Figure 4.8
The target specifications. Like the other information systems, this one is easily encoded with a spreadsheet as a simple extension to the list of specifications.

Step 4: Reflect on the Results and the Process

The team may require some iteration to agree on the targets. Reflection after each iteration helps to ensure that the results are consistent with the goals of the project. Questions to consider include:

- Are members of the team "gaming"? For example, is the key marketing representative insisting that an aggressive value is required for a particular metric in the hopes that by setting a high goal, the team will actually achieve more than if his or her true, and more lenient, beliefs were expressed?
- Should the team consider offering multiple products or at least multiple options for the product in order to best match the particular needs of more than one market segment, or will one "average" product suffice?
- Are any specifications missing? Do the specifications reflect the characteristics that will dictate commercial success?

Once the targets have been set, the team can proceed to generate solution concepts. The target specifications then can be used to help the team select a concept and will help the team know when a concept is commercially viable. (See Chapter 6, Concept Generation, and Chapter 7, Concept Selection.)

4.4 | SETTING THE FINAL SPECIFICATIONS

As the team finalizes the choice of a concept and prepares for subsequent design and development, the specifications are revisited. Specifications which originally were only targets expressed as broad ranges of values are now refined and made more precise.

Finalizing the specifications is difficult because of trade-offs—inverse relationships between two specifications that are inherent in the selected product concept. Trade-offs frequently occur between different technical performance metrics and almost always occur between technical performance metrics and cost. For example, one trade-off is between brake mounting stiffness and mass of the fork. Because of the basic mechanics of the fork structure, these specifications are inversely related, assuming other factors are held constant. Another trade-off is between cost and mass. For a given concept, the team may be able to reduce the mass of the fork by making some parts out of titanium instead of steel. Unfortunately, decreasing the mass in this way will most likely increase the manufacturing cost of the product. The difficult part of re-

fining the specifications is choosing how such trade-offs will be resolved.

Here, we propose a five-step process:

1. Develop technical models of the product.
2. Develop a cost model of the product.
3. Refine the specifications, making trade-offs where necessary.
4. Flow down the specifications as appropriate.
5. Reflect on the results and the process.

Step 1: Develop Technical Models of the Product

A *technical model* of the product is a tool for predicting the values of the metrics for a particular set of design decisions. We intend the term *models* to refer to both analytical and physical approximations of the product. (See Chapter 12, Prototyping, for further discussion of such models.)

At this point, the team had chosen an oil-damped coil spring concept for the suspension fork. The design decisions facing the team included details such as the materials for the structural components, the orifice diameter and oil viscosity for the damper, and the spring constant. Three models linking such design decisions to the performance metrics are shown in conceptual form in Figure 4.9. Such models can be used to predict the product's performance along a number of dimensions. The inputs to these models are the independent design variables associated with the product concept, such as oil viscosity, orifice diameter, spring constant, and geometry. The outputs of the model are the values of the metrics, such as attenuation, stiffness, and fatigue life.

Ideally, the team will be able to accurately model the product analytically, perhaps by implementing the model equations in a spreadsheet or computer simulation. Such a model allows the team to predict rapidly what type of performance can be expected from a particular choice of design variables, without costly physical experimentation. In most cases, such analytical models will be available for only a small subset of the metrics. For example, the team was able to model attenuation analytically, based on the engineers' knowledge of dynamic systems.

Several independent models, each corresponding to a subset of the metrics, may be more manageable than one large integrated model. For example, the team developed a separate analytical model for the brake mounting stiffness that was completely independent of the dynamic model used to predict vibration attenuation. In some cases, no analytical models will be available at all. For example, the team was not able to model analytically the fatigue performance of the suspension, so physical models were built and tested. It is generally necessary to actually build a variety of different physical mock-ups or prototypes in order to explore the implications of several combinations of design variables. When physical models must be constructed, it is often useful to use design-of-experiments (DOE) techniques, which can minimize the number of experiments required to explore the design space (Box et al., 1978; Phadke, 1989).

Armed with these technical models, the team can predict whether or not any particular set of specifications (such as the ideal target values) is technically feasible by exploring different combinations of design variables. This type of modeling and analysis prevents the team from setting a combination of specifications that cannot be achieved using the available latitude in the product concept.

Note that a technical model is almost always unique to a particular product concept. One of the models illustrated in Figure 4.9 is for an oil-damped suspension system; the model would be substantially different if the team had selected a concept employing a rubber suspension element. Thus, the modeling step can only be performed after the concept has been chosen.

Step 2: Develop a Cost Model of the Product

The goal of this step of the process is to make sure that the product can be produced at the *target cost*. The target cost is the manufacturing cost at which the company and its distribution partners can make adequate profits while still offering the product to the end customer at a competitive price. The appendix to this chapter provides an explanation of target costing. It is at this point that the team attempts to discover, for example, how much it will have to sacrifice in manufacturing cost to save 50 grams of mass.

For most products, the first estimates of manufacturing costs are completed by drafting a *bill of materials* (a list of all the parts) and estimating a purchase price or fabrication cost for each part. At this point in the development process the team does not generally know all of the components that will be in the product, but the team nevertheless makes an attempt to list the components it expects will be required. While early estimates generally focus on the cost of components, the team will usually make a rough estimate of assembly and other manufacturing costs (e.g., overhead) at this point as well. Efforts to develop these early cost estimates involve soliciting cost estimates from vendors and estimating the production costs of the components the firm will make itself. This process is often facilitated by a pur-

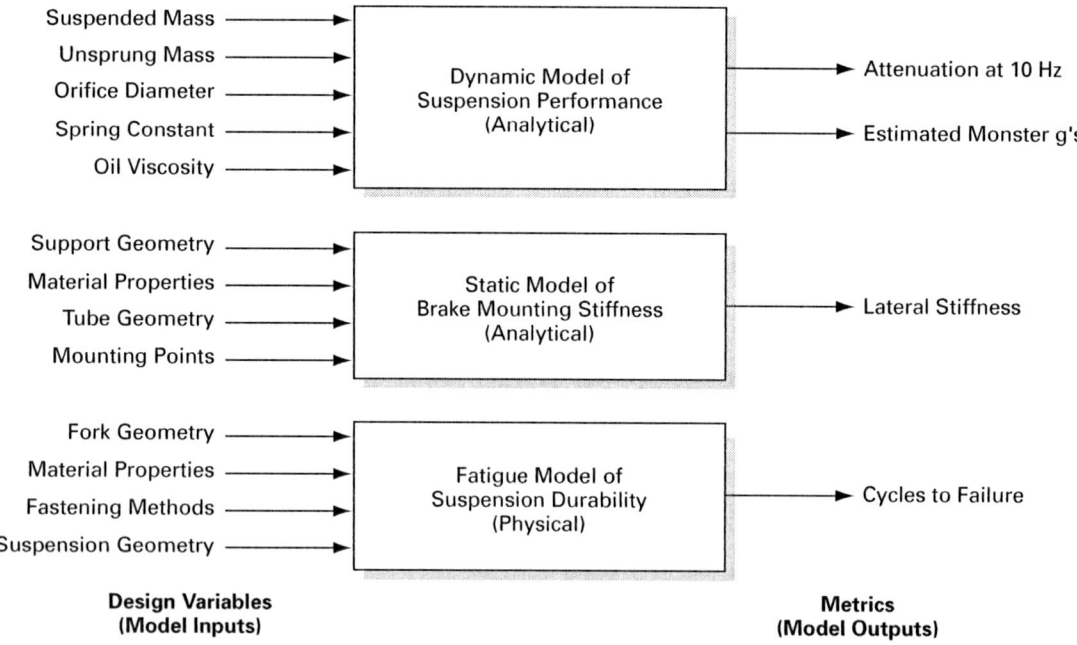

Figure 4.9

Models used to assess technical feasibility. Technical models may be analytical or physical approximations of the product concept.

chasing expert and a production engineer. A bill-of-materials cost model is shown in Figure 4.10 for the suspension fork. (See Chapter 11, Design for Manufacturing, for more details on estimating manufacturing cost.)

A useful way to record cost information is to list figures for the high and low estimates of each item. This helps the team to understand the range of uncertainty in the estimates. The bill of materials is typically used iteratively: the team performs a "what if" cost analysis for a set of design decisions and then revises these decisions based on what it learns. The bill of materials is itself a kind of performance model, but instead of predicting the value of a technical performance metric, it predicts cost performance. The bill of materials remains useful throughout the development process and is updated regularly (as fre-

quently as once each week) to reflect the current status of the estimated manufacturing cost.

At this point in the development process, teams developing complex products containing hundreds or thousands of parts will not generally be able to include every part in the bill of materials. Instead, the team will list the major components and sub-systems and place bounds on their costs based on past experience or on the judgment of suppliers.

Step 3: Refine the Specifications, Making Trade-Offs Where Necessary

Once the team has constructed technical performance models where possible and constructed a preliminary cost model, these tools can be used to develop final specifica-

Component	Qty/Fork	High ($ ea.)	Low ($ ea.)	High Total ($/fork)	Low Total ($/fork)
Steer tube	1	2.50	2.00	2.50	2.00
Crown	1	4.00	3.00	4.00	3.00
Boot	2	1.00	0.75	2.00	1.50
Lower tube	2	3.00	2.00	6.00	4.00
Lower tube top cover	2	2.00	1.50	4.00	3.00
Main lip seal	2	1.50	1.40	3.00	2.80
Slide bushing	4	0.20	0.18	0.80	0.72
Slide bushing spacer	2	0.50	0.40	1.00	0.80
Lower tube plug	2	0.50	0.35	1.00	0.70
Upper tube	2	5.50	4.00	11.00	8.00
Upper tube top cap	2	3.00	2.50	6.00	5.00
Upper tube adjustment knob	2	2.00	1.75	4.00	3.50
Adjustment shaft	2	4.00	3.00	8.00	6.00
Spring	2	3.00	2.50	6.00	5.00
Upper tube orifice cap	1	3.00	2.25	3.00	2.25
Orifice springs	4	0.50	0.40	2.00	1.60
Brake studs	2	0.40	0.35	0.80	0.70
Brake brace bolt	2	0.25	0.20	0.50	0.40
Brake brace	1	5.00	3.50	5.00	3.50
Oil (liters)	0.1	2.50	2.00	0.25	0.20
Misc. snap rings, o-rings	10	0.15	0.10	1.50	1.00
Decals	4	0.25	0.15	1.00	0.60
Assembly at $20/hr		30 min	20 min	10.00	6.67

Figure 4.10
A bill of materials with cost estimates. This simple cost model allows early cost estimates to facilitate realistic trade-offs in the product specifications.

tions. Finalizing specifications can be accomplished in a group session in which feasible combinations of values are determined through the use of the technical models and then the cost implications are explored. In an iterative fashion, the team converges on the specifications which will most favorably position the product relative to the competition, will best satisfy the customer needs, and will assure adequate profits.

One important tool for supporting this decision-making process is the *competitive map*. An example competitive map is shown in Figure 4.11. This map is simply a scatter plot of the competitive products along two dimensions selected from the set of metrics. The map displayed in Figure 4.11 shows estimated manufacturing cost versus g's on the Monster test. The regions defined by the marginal and ideal values of the specifications are shown on the map. This map is particularly useful in showing that all of the high-performance suspensions (low Monster scores) have high estimated manufacturing costs. Armed with technical performance models and a cost model, the team can assess whether or not it will be able to "beat the trade-off" exhibited in the competitive map.

These maps can be constructed directly from the data contained in the competitive benchmarking chart using the plotting feature of the spreadsheet software. Generally the team will prepare three or four such maps corresponding to a handful of critical metrics. Additional maps may be created as needed to support subsequent decision making.

The competitive map is used to position the new product relative to the competition. Using the technical and cost models of the product and the competitive maps, the team can refine the specifications in order to both satisfy the inherent constraints of the product concept and make the trade-offs in a way that will provide a performance advantage relative to the competitive products. The final specifications for the suspension fork are shown in Figure 4.12.

For relatively mature product categories in which competition is based on performance relative to a handful of well-understood performance metrics, *conjoint analysis* may be useful in refining product specifications. Conjoint analysis uses customer survey data to construct a model of customer preference. Essentially each respondent in a sample of potential customers is repeatedly asked to evaluate hypothetical products characterized by a set of attributes. These attributes must generally be metrics that are easily understood by customers (e.g., fuel economy and price for automobiles). Subjective attributes (e.g., styling) can be represented graphically. The hypothetical products are

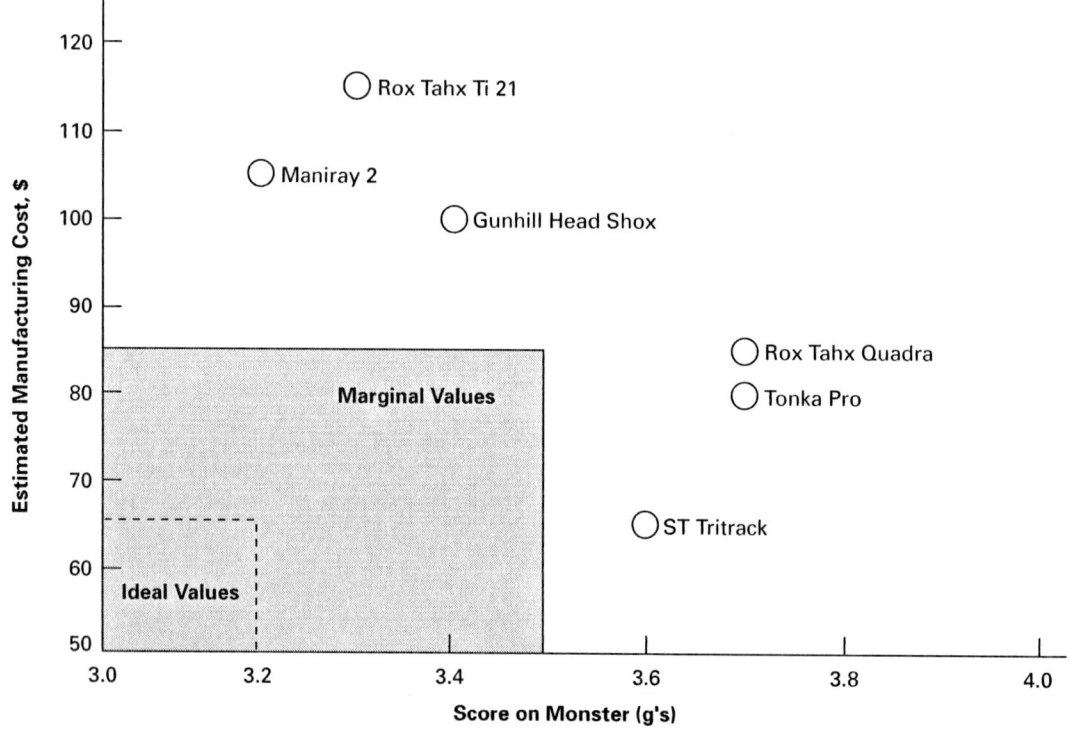

Figure 4.11

A competitive map showing estimated manufacturing cost versus score on the "Monster" test.

constructed using the statistical techniques of experimental design. Using customer responses, conjoint analysis infers the relative importance of each attribute to the customer. These data can then be used to predict which product a customer would choose when offered a hypothetical set of alternatives. By using these predictions for all of the customers in a sample, the market share of each product in the set of alternatives can be forecast. Using this approach, the specification values that maximize market share can be estimated. The details of conjoint analysis are fairly straightforward, but beyond the scope of this chapter. Relevant references are listed at the end of the chapter.

Step 4: Flow Down the Specifications as Appropriate

This chapter focuses on the specifications for a relatively simple component designed by a single, relatively small development team. Establishing specifications takes on additional importance and is substantially more challenging when developing a highly complex product consisting of multiple subsystems designed by multiple development teams. In such a context, specifications are used to define the development objectives of each of the subsystems as well as for the product as a whole. The challenge in this case is to *flow down* the overall specifications to specifications for each subsystem. For example, the overall specifi-

No.	Metric	Unit	Value
1	Attenuation from dropout to handlebar at 10Hz	dB	>12
2	Spring preload	N	600-650
3	Maximum value from the Monster	g	<3.4
4	Minimum descent time on test track	s	<11.5
5	Damping coefficient adjustment range	N-s/m	>100
6	Maximum travel (26-in. wheel)	mm	43
7	Rake offset	mm	38
8	Lateral stiffness at the tip	kN/m	>75
9	Total mass	kg	<1.4
10	Lateral stiffness at brake pivots	kN/m	>425
11	Headset sizes	in.	1.000 1.125
12	Steertube length	mm	150 170 190 210 230
13	Wheel sizes	List	26 in.
14	Maximum tire width	in.	>1.75
15	Time to assemble to frame	s	<45
16	Fender compatibility	List	Zefal
17	Instills pride	Subj.	>4
18	Unit manufacturing cost	US$	<80
19	Time in spray chamber without water entry	s	>3600
20	Cycles in mud chamber without contamination	k-cycles	>25
21	Time to disassemble/assemble for maintenance	s	<200
22	Special tools required for maintenance	List	Hex
23	UV test duration to degrade rubber parts	Hours	>450
24	Monster cycles to failure	Cycles	>500k
25	Japan Industrial Standards test	Binary	Pass
26	Bending strength (frontal loading)	kN	>10.0

Figure 4.12
The final specifications.

cations for an automobile contain metrics like fuel economy, 0-100 kilometer/hour acceleration time, and turning radius. However, specifications must also be created for the several dozen major subsystems that make up the automobile, including the body, engine, transmission, braking system, and suspension. The specifications for the engine include metrics like peak power, peak torque, and fuel consumption at peak efficiency. One challenge in the flow-down process is to ensure that the subsystem specifications in fact reflect the overall product specifications—that if specifications for the subsystems are achieved, the overall product specifications will be achieved. A second challenge is to ensure that certain specifications for different subsystems are equally difficult to meet. That is, for example, that the mass specification for the engine is not inordinately more difficult to meet than is the mass specification for the body. Otherwise, the cost of the product will likely be higher than necessary.

Some overall component specifications can be established through *budget allocations*. For example, specifications for manufacturing cost, mass, and power consumption can be allocated to subsystems with the confidence that the overall cost, mass, and power consumption of the product will simply be the sum of these quantities for each subsystem. To some extent, geometric volume can be allocated this way as well. Other component specifications must be established through a more complex understanding of how subsystem performance relates to overall product performance. For example, fuel efficiency is a relatively complex function of vehicle mass, rolling resistance, aerodynamic drag coefficient, frontal area, and engine efficiency. Establishing specifications for the body, tires, and engine requires a model of how these variables relate to overall fuel efficiency.

A comprehensive treatment of flowing down specifications for complex products is beyond the scope of this chapter, and in fact is a major focus of the field of *systems engineering*. We refer the reader to several useful books on this subject in the reference list.

Step 5: Reflect on the Results and the Process

As always, the final step in the method is to reflect on the outcome and the process. Some questions the team may want to consider are:

- Is the product a winner? The product concept should allow the team to actually set the specifications so that the product will meet the customer needs and excel competitively. If not, then the team should return to the concept generation and selection phase or abandon the project.

- How much uncertainty is there in the technical and cost models? If competitive success is dictated by metrics around which much uncertainty remains, the team may wish to refine the technical or cost models in order to increase confidence in meeting the specifications.

- Is the concept chosen by the team best suited to the target market, or could it be best applied in another market (say, the low end or high end instead of the middle)? The selected concept may actually be too good. If the team has generated a concept that is dramatically superior to the competitive products, it may wish to consider employing the concept in a more demanding, and potentially more profitable, market segment.

- Should the firm initiate a formal effort to develop better technical models of some aspect of the product's performance for future use? Sometimes the team will discover that it does not really understand the underlying product technology well enough to create useful performance models. In such circumstances, an engineering effort to develop better understanding and models may be useful in subsequent development projects.

4.5 | SUMMARY

Customer needs are generally expressed in the "language of the customer." In order to provide specific guidance about how to design and engineer a product, development teams establish a set of specifications, which spell out in precise, measurable detail what the product has to do in order to be commercially successful. The specifications must reflect the customer needs, differentiate the product from the competitive products, and be technically and economically realizable.

- Specifications are typically established at least twice. Immediately after identifying the customer needs, the team sets *target specifications*. After concept selection and testing, the team develops *final specifications*.
- Target specifications represent the hopes and aspirations of the team, but they are established before the team knows the constraints the product technology will place on what can be achieved. The team's efforts may fail to meet some of these specifications and may exceed others, depending on the details of the product concept the team eventually selects.
- The process of establishing the target specifications contains four steps:

1. Prepare the list of metrics, using the needs-metrics matrix, if necessary.
2. Collect the competitive benchmarking information.
3. Set *ideal* and *marginally acceptable* target values for each metric.
4. Reflect on the results and the process.

- Final specifications are developed by assessing the actual technological constraints and the expected production costs using analytical and physical models. During this refinement phase the team must make difficult trade-offs among various desirable characteristics of the product.
- The five-step process for refining the specifications is:
    1. Develop technical models of the product.
    2. Develop a cost model of the product.
    3. Refine the specifications, making trade-offs where necessary.
    4. Flow down the specifications as appropriate.
    5. Reflect on the results and the process.
- The specifications process is facilitated by several simple *information systems* which can easily be created using conventional spreadsheet software. Tools such as the list of metrics, the needs-metrics matrix, the competitive benchmarking charts, and the competitive maps all support the team's decision making by providing the team with a way to represent and discuss the specifications.
- Because of the need to utilize the best possible knowledge of the market, the customers, the core product technology, and the cost implications of design alternatives, the specifications process requires active participation from team members representing the marketing, design, and manufacturing functions of the enterprise.

4.6 | REFERENCES AND BIBLIOGRAPHY

Many current resources are available on the Internet via

www.ulrich-eppinger.net

The process of translating customer needs into a set of specifications is also accomplished by the Quality Function Deployment (QFD) method. The key ideas behind QFD and the House of Quality are clearly presented by Hauser and Clausing in a popular article.

Hauser, John, and Don Clausing, "The House of Quality," *Harvard Business Review,* Vol. 66, No. 3, May-June 1988, pp.63-73.

Urban and Hauser present several techniques for selecting combinations of product attributes in order to maximize customer satisfaction. Some of these techniques can serve as powerful analytical support for the general method described in this chapter.

Urban, Glen, and John Hauser, *Design and Marketing of New Products,* second edition, Prentice Hall, Englewood Cliffs, NJ, 1993.

Ramaswamy and Ulrich treat the use of engineering models in setting specifications in detail. They also identify some of the weaknesses in the conventional House of Quality method.

Ramaswamy, Rajan, and Karl Ulrich, "Augmenting the House of Quality with Engineering Models," *Research in Engineering Design,* Vol. 5, 1994, pp.70-79.

Most marketing research textbooks discuss conjoint analysis. Here are two references.

Conjoint Analysis: A Guide for Designing and Interpreting Conjoint Studies, American Marketing Association, June 1992.

Aaker, David A., V. Kumar, and George S. Day, *Marketing Research,* sixth edition, John Wiley & Sons, New York, 1997.

Box et al. provide detailed explanation of the use of statistical methods, including fractional factorial methods, for experimentation. Phadke describes Taguchi's methods for design of experiments.

Box, George E. P., J. Stuart Hunter, and William G. Hunter, *Statistics for Experimenters: An Introduction to Design, Data Analysis, and Model Building,* John Wiley and Sons, New York, 1978.

Phadke, Madhav S., *Quality Engineering Using Robust Design,* Prentice Hall, Englewood Cliffs, NJ, 1989.

Systems engineering and the flow down of specifications are treated comprehensively in the following books.

Hatley, Derek J., and Imtiaz A. Pirbhai, *Strategies for Real-Time System Specification,* Dorset House, 1998.

Rechtin, Eberhardt, and Mark W. Maier, *The Art of Systems Architecting,* CRC Press, 1996.

More detail on the use of target costing is available in this article by Cooper and Slagmulder.

Cooper, Robin, and Regine Slagmulder, "Develop Profitable New Products with Target Costing," *Sloan Management Review,* Vol. 40, No. 4, Summer 1999, pp.23-33.

Exercises

1. List a set of metrics corresponding to the need that a pen write smoothly.

2. Devise a metric and a corresponding test for the need that a roofing material last many years.

3. Some of the same metrics seem to be involved in trade-offs for many different products. Which metrics are these?

Thought Questions

1. How might you establish precise and measurable specifications for intangible needs such as "the front suspension looks great"?

2. Why are some customer needs difficult to map to a single metric?

3. How might you explain a situation in which customers' perceptions of the competitive products (as in Figure 4.7) are not consistent with the values of the metrics for those same products (as in Figure 4.6)?

4. Can poor performance relative to one specification always be compensated for by high performance on other specifications? If so, how can there ever really be a "marginally acceptable" value for a metric?

5. Why should independent design variables not be used as metrics?

Appendix: Target Costing

Target costing is a simple idea: set the value of the manufacturing cost specification based on the price the company hopes the end user will pay for the product and on the profit margins that are required for each stage in the distribution channel. For example, assume Specialized wishes to sell its suspension fork to its customers through bicycle shops. If the price it expected the customer to pay were $250 and if bicycle shops normally expect a gross profit margin of 45 percent on components, then Specialized would have to sell its fork to bicycle shops for $(1 - 0.45) \times 250 = \$137.50$. If Specialized wishes to earn a gross margin of at least 40 percent on its components, then its unit manufacturing cost must be less than $(1 - 0.40) \times 137.50 = \$82.50$.

Target costing is the reverse of the *cost-plus* approach to pricing. The cost-plus approach begins with what the firm expects its manufacturing costs to be and then sets its prices by adding its expected profit margin to the cost. This approach ignores the realities of competitive markets, in which prices are driven by market and customer factors. Target costing is a mechanism for ensuring that specifications are set in a way that allows the product to be competitively priced in the marketplace.

Some products are sold directly by a manufacturer to end users of the product. Frequently, products are distributed through one or more intermediate stages, such as distributors and retailers. Figure 4.A1 provides some approximate values of target gross profit margins for different product categories.

Let M be the gross profit *margin* of a stage in the distribution channel.

$$M = \frac{(P - C)}{P}$$

where P is the price this stage charges its customers and C is the cost this stage pays for the product it sells. (Note that *mark-up* is similar to margin, but is defined slightly differently as $P/C - 1$, so that a margin of 50 percent is equivalent to a mark-up of 100 percent.)

Target cost, C, is given by the following expression:

$$C = P \prod_{i=1}^{n} (1 - M_i)$$

where P is the price paid by the end user, n is the number of stages in the distribution channel, and M_i is the margin of the ith stage.

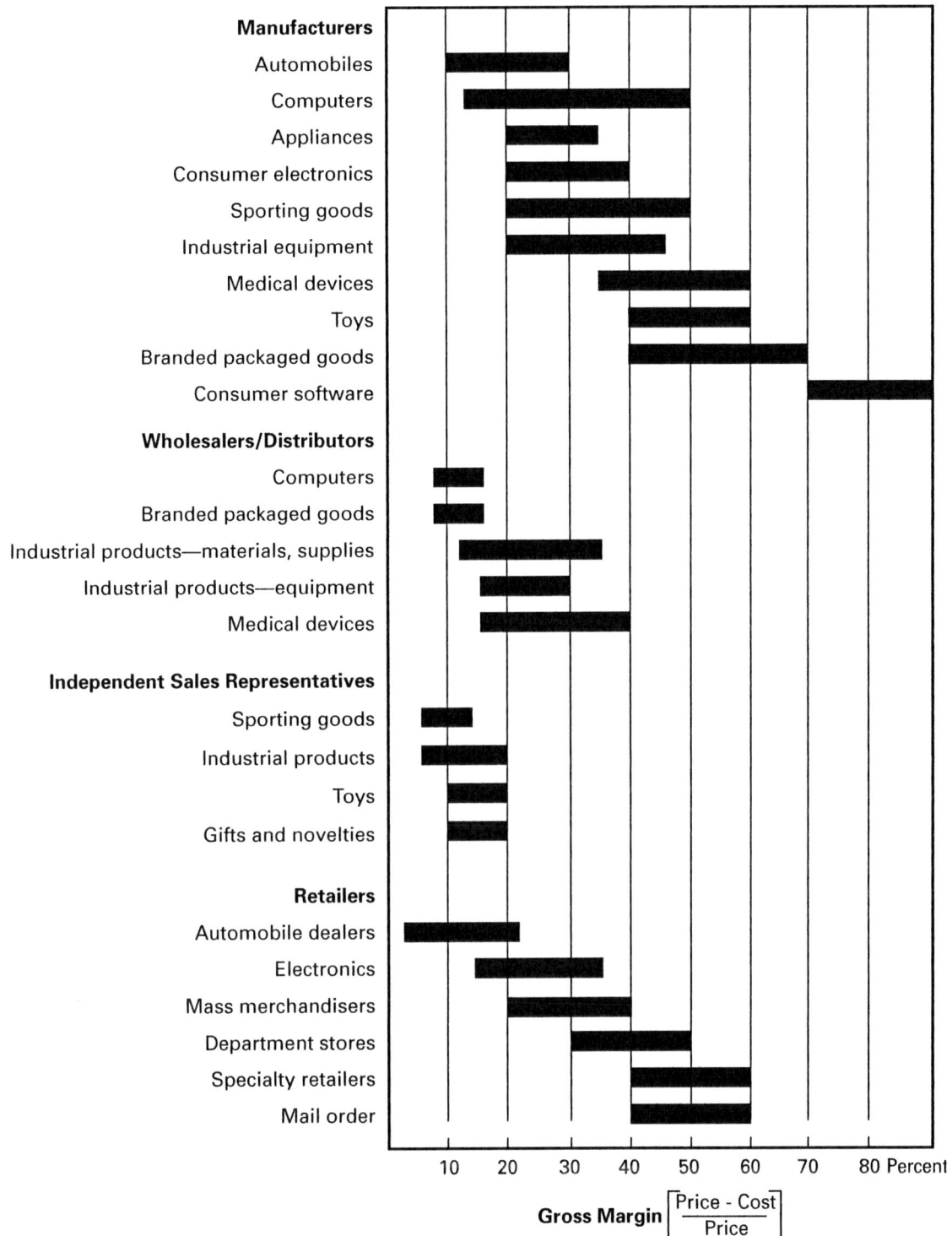

Figure 4.A1
Approximate margins for manufacturers, wholesalers, distributors, sales representatives, and retailers. Note that these values are quite approximate. Actual margins depend on many idiosyncratic factors, including competitive intensity, the volume of units sold, and the level of customer support required.

Example

Assume the end user price, P, equals $250.

If the product is sold directly to the end user by the manufacturer, and the desired gross profit margin of the manufacturer, M_m, equals 0.40, then the target cost is:

$$C = P(1 - M_m) = \$250(1 - 0.40) = \$150$$

If the product is sold through a retailer, and the desired gross profit margin for the retailer, M_r, equals 0.45, then:

$$C = P(1 - M_m)(1 - M_r)$$
$$= \$250(1 - 0.40)(1 - 0.45) = \$82.50$$

If the product is sold through a distributor and a retailer, and the desired gross profit margin for the distributor, M_d, equals 0.20, then:

$$C = P(1 - M_m)(1 - M_d)(1 - M_r)$$
$$= \$250(1 - 0.40)(1 - 0.20)(1 - 0.45) = \$66.00$$

This chapter was developed in
collaboration with Gavin Zau.

Concept Generation

5

The president of a tool manufacturing company commissioned a team to develop a new hand-held nailer for the high-end consumer market. One of several currently available pneumatic tools for the professional market is shown in Figure 5.1. The mission of the team was to consider broadly alternative product concepts, assuming only that the tool would employ conventional nails as the basic fastening technology. After identifying a set of customer needs and establishing target product specifications, the team faced the following questions:

- What existing solution concepts, if any, could be successfully adapted for this application?
- What new concepts might satisfy the established needs and specifications?
- What methods can be used to facilitate the concept generation process?

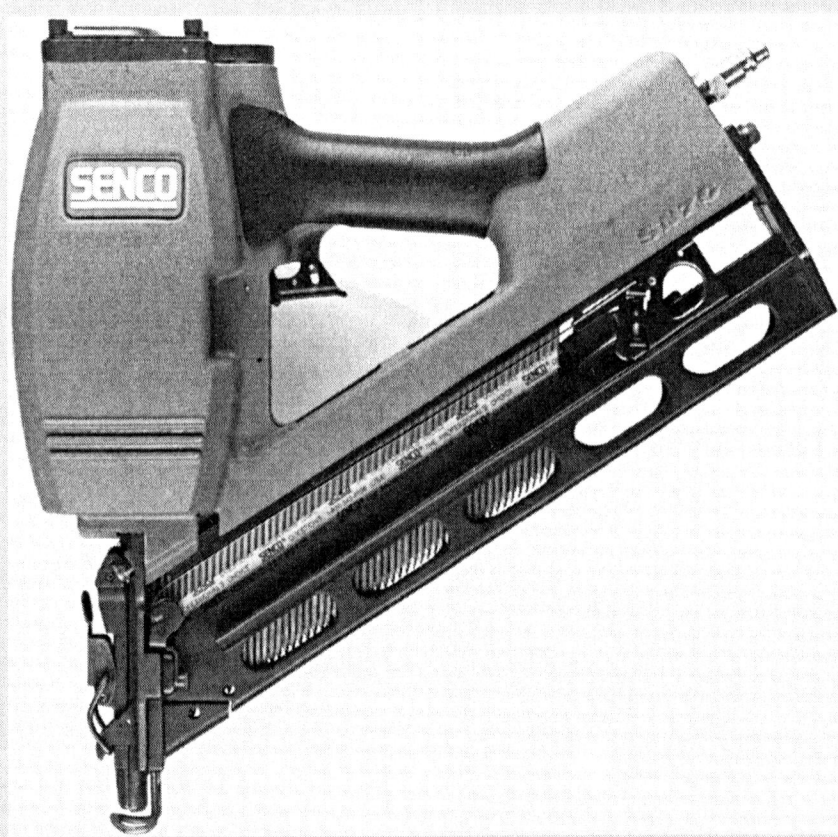

Figure 5.1
An existing hand-held nailer for the professional market. (Courtesy of Senco Products.)

5.1 | THE ACTIVITY OF CONCEPT GENERATION

A product concept is an approximate description of the technology, working principles, and form of the product. It is a concise description of how the product will satisfy the customer needs. A concept is usually expressed as a sketch or as a rough three-dimensional model and is often accompanied by a brief textual description. The degree to which a product satisfies customers and can be successfully commercialized depends to a large measure on the quality of the underlying concept. A good concept is sometimes poorly implemented in subsequent development phases, but a poor concept can rarely be manipulated to achieve commercial success. Fortunately, concept generation is relatively inexpensive and can be done relatively quickly in comparison to the rest of the development process. For example, concept generation had typically consumed less than 5 percent of the budget and 15 percent of the development time in previous nailer development efforts. Because the concept generation activity is not costly, there is no excuse for a lack of diligence and care in executing a sound concept generation method.

The concept generation process begins with a set of customer needs and target specifications and results in a set of product concepts from which the team will make a final selection. The relation of concept generation to the other concept development activities is shown in Figure 5.2. In most cases, an effective development team will generate hundreds of concepts, of which 5 to 20 will merit serious consideration during the concept selection activity.

Good concept generation leaves the team with confidence that the full space of alternatives has been explored. Thorough exploration of alternatives early in the development process greatly reduces the likelihood that the team will stumble upon a superior concept late in the development process or that a competitor will introduce a product with dramatically better performance than the product under development.

Structured Approaches Reduce the Likelihood of Costly Problems

Common dysfunctions exhibited by development teams during concept generation include:

- Consideration of only one or two alternatives, often proposed by the most assertive members of the team.
- Failure to consider carefully the usefulness of concepts employed by other firms in related and unrelated products.
- Involvement of only one or two people in the process, resulting in lack of confidence and commitment by the rest of the team.
- Ineffective integration of promising partial solutions.
- Failure to consider entire categories of solutions.

A structured approach to concept generation reduces the incidence of these problems by encouraging the gathering of information from many disparate information sources, by guiding the team in the thorough exploration of alternatives, and by providing a mechanism for integrating partial solutions. A structured method also provides a step-by-step procedure for those members of the team who may be less experienced in design-intensive activities, allowing them to participate actively in the process.

A Five-Step Method

This chapter presents a five-step concept generation method. The method, outlined in Figure 5.3, breaks a complex problem into simpler subproblems. Solution concepts are then identified for the subproblems by external and internal search procedures. Classification trees and concept combination tables are then used to systematically explore the space of solution concepts and to integrate the subproblem solutions into a total solution. Finally, the team takes a step back to reflect on the validity and applicability of the results, as well as on the process used.

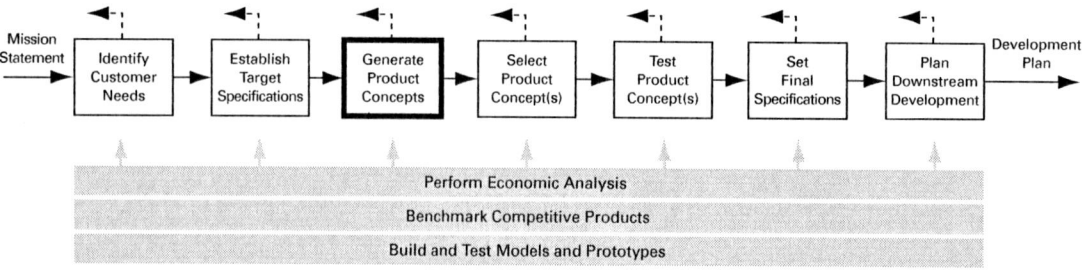

Figure 5.2
Concept generation is an integral part of the concept development phase.

This chapter will follow the recommended method and will describe each of the five steps in detail. Although we present the method in a linear sequence, concept generation is almost always iterative. Like our other development methods, these steps are intended to be a baseline from which product development teams can develop and refine their own unique problem-solving style.

Our presentation of the method is focused primarily on the overall concept for a new product; however, the method can and should be used at several different points in the development process. The process is useful not only for overall product concepts but also for concepts for subsystems and specific components as well. Also note that while the example in this chapter involves a relatively technical

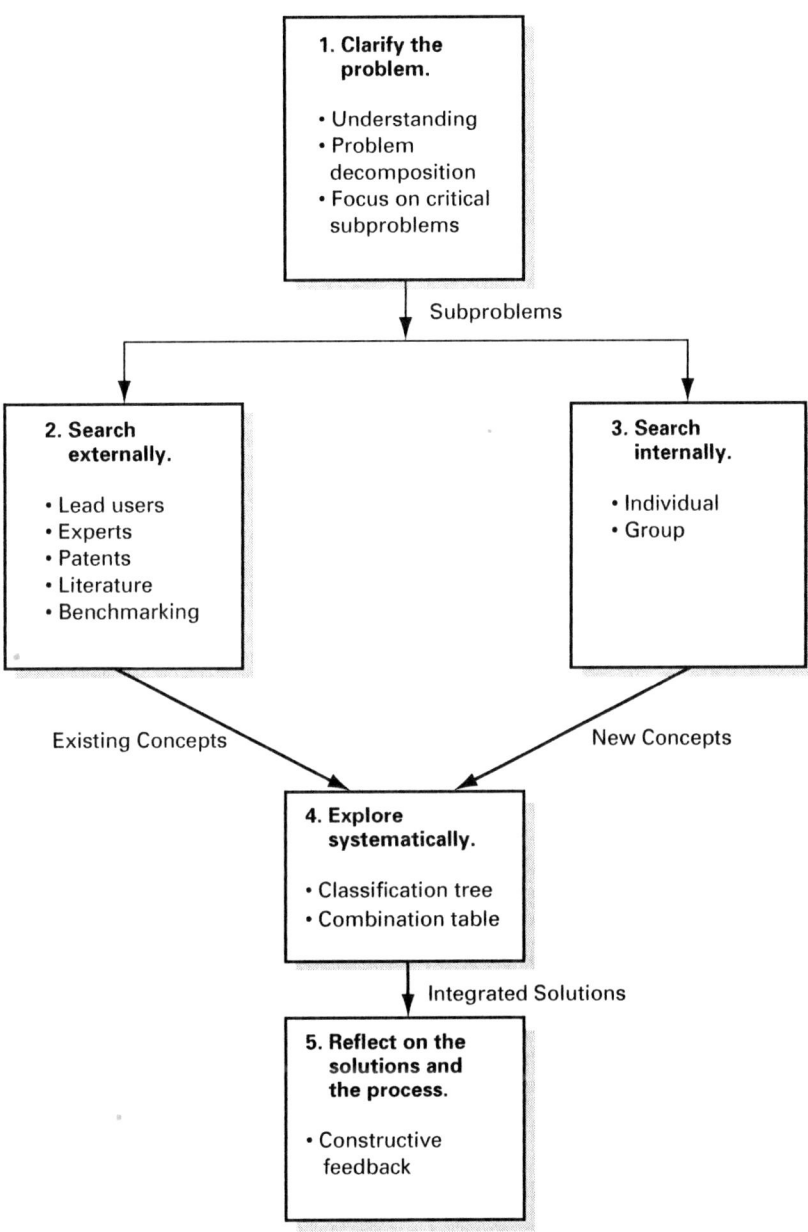

Figure 5.3
The five-step concept generation method.

product, the same basic approach can be applied to nearly any product.

5.2 | STEP 1: CLARIFY THE PROBLEM

Clarifying the problem consists of developing a general understanding and then breaking the problem down into subproblems if necessary.

The mission statement for the project, the customer needs list, and the preliminary product specification are the ideal inputs to the concept generation process, although often these pieces of information are still being refined as the concept generation phase begins. Ideally the team has been involved in both the identification of the customer needs and in the setting of the target product specifications. Those members of the team who were not involved in these preceding steps should become familiar with the processes used and their results before concept generation activities begin. (See Chapter 4, Identifying Customer Needs, and Chapter 5, Product Specifications.)

As stated before, the challenge was to "design a better hand-held nailer." The scope of the design problem could have been defined more generally (e.g., "fasten wood together") or more specifically (e.g., "improve the speed of the existing pneumatic tool concept"). Some of the assumptions in the team's mission statement were:

- The nailer will use nails (as opposed to adhesives, screws, etc.).
- The nailer will be compatible with nail magazines on existing tools.
- The nailer will nail into wood.
- The nailer will be hand held.

Based on the assumptions, the team had identified the customer needs for a hand-held nailer. These included:

- The nailer inserts nails in rapid succession.
- The nailer fits into tight spaces.
- The nailer is lightweight.
- The nailer has no noticeable nailing delay after tripping the tool.

The team gathered supplemental information to clarify and quantify the needs, such as the approximate energy and speed of the nailing. These basic needs were subsequently translated into target product specifications. The target specifications included the following:

- Nail lengths from 50 millimeters to 75 millimeters.
- Maximum nailing energy of 80 joules per nail.

- Nailing forces of up to 2,000 newtons.
- Peak nailing rate of one nail per second.
- Average nailing rate of four nails per minute.
- Ability to insert nails between standard stud/joists (368 millimeter opening).
- Tool mass less than 4 kilograms.
- Maximum trigger delay of 0.25 seconds.

Decompose a Complex Problem into Simpler Subproblems

Many design challenges are too complex to solve as a single problem and can be usefully divided into several simpler subproblems. For example, the design of a complex product like a document copier can be thought of as a collection of more focused design problems, including for example, the design of a document handler, the design of a paper feeder, the design of a printing device, and the design of an image capture device. In some cases, however, the design problem cannot readily be divided into subproblems. For example, the problem of designing a paper clip may be hard to divide into subproblems. As a general rule, we feel that teams should attempt to decompose design problems, but should be aware that such a decomposition may not be very useful for products with extremely simple functions.

Dividing a problem into simpler subproblems is called *problem decomposition*. There are many schemes by which a problem can be decomposed. Here we demonstrate a *functional* decomposition and also list several other approaches that are frequently useful.

The first step in decomposing a problem functionally is to represent it as a single *black box* operating on material, energy, and signal flows, as shown in Figure 5.4(a). Thin solid lines denote the transfer and conversion of energy, thick solid lines signify the movement of material within the system, and dashed lines represent the flows of control and feedback signals within the system. This black box represents the overall function of the product.

The next step in functional decomposition is to divide the single black box into subfunctions to create a more specific description of what the elements of the product might do in order to implement the overall function of the product. Each subfunction can generally be further divided into even simpler subfunctions. The division process is repeated until the team members agree that each subfunction is simple enough to work with. A good rule of thumb is to create between 3 and 10 subfunctions in the diagram. The end result, shown in Figure 5.4(b), is a function diagram containing subfunctions connected by energy, material, and signal flows.

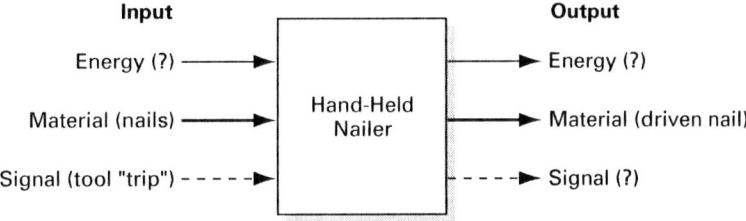

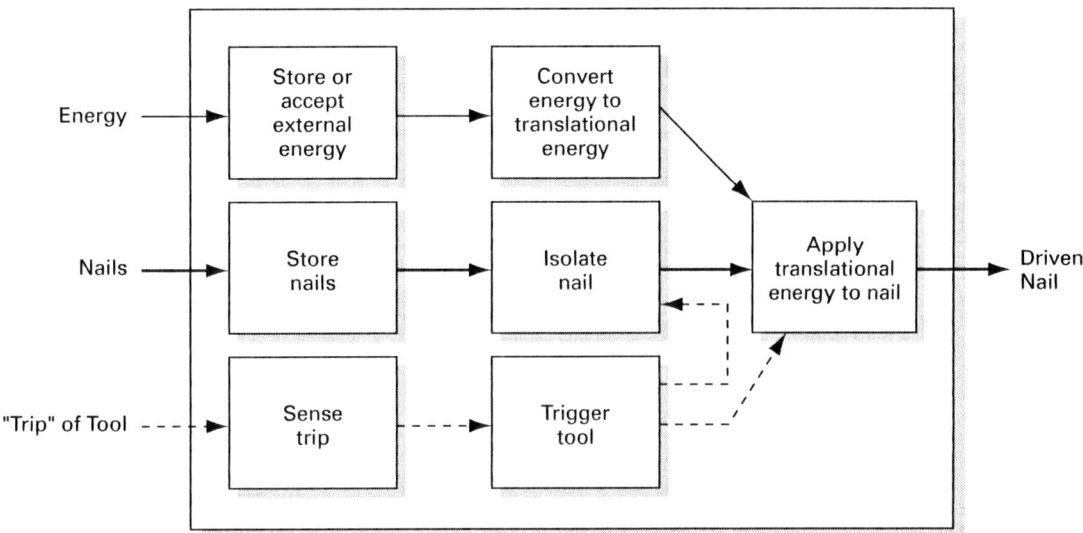

Figure 5.4
Function diagram of a hand-held nailer arising from a functional decomposition: (a)overall "black box"; (b)refinement showing subfunctions.

Note that at this stage the goal is to describe the functional elements of the product without implying a specific technological working principle for the product concept. For example, Figure 5.4(b) includes the subfunction "isolate nail." This subfunction is expressed in such a way that it does not imply any particular physical solution concept, such as indexing the "stick" of nails into a slot or breaking a nail sideways off of the stick. The team should consider each subfunction in turn and ask whether it is expressed in a way that does not imply a particular physical solution principle.

There is no single correct way of creating a function diagram and no single correct functional decomposition of a product. A helpful way to create the diagram is to quickly create several drafts and then work to refine them into a single diagram that the team is comfortable with. Some useful techniques for getting started are:

- Create a function diagram of an existing product.
- Create a function diagram based on an arbitrary product concept already generated by the team or based on a known subfunction technology. Be sure to generalize the diagram to the appropriate level of abstraction.
- Follow one of the flows (e.g., material) and determine what operations are required. The details of the other flows can be derived by thinking about their connections to the initial flow.

Note that the function diagram is typically not unique. In particular, subfunctions can often be ordered in different ways to produce different function diagrams. Also note that in some applications the material, energy, and signal flows are difficult to identify. In these cases, a simple list of the subfunctions of the product, without connections between them, is often sufficient.

Functional decomposition is most applicable to technical products, but it can also be applied to simple and apparently nontechnical products. For example, an ice cream scoop has material flow of ice cream being separated, formed, transported, and deposited. These subfunctions could form the basis of a problem decomposition.

Functional decomposition is only one of several possible ways to divide a problem into simpler subproblems. Two other approaches are described below:

- *Decomposition by sequence of user actions:* For example, the nailer problem might be broken down into three user actions: moving the tool to the gross nailing position, positioning the tool precisely, triggering the tool. This approach is often useful for products with very simple technical functions involving a lot of user interaction.
- *Decomposition by key customer needs:* For the nailer, this decomposition might include the following subproblems: fires nails in rapid succession, fits in tight places, and has a large nail capacity. This approach is often useful for products in which form, and not working principles or technology, is the primary problem. Examples of such products include toothbrushes (assuming the basic brush concept is retained) and storage containers.

Focus Initial Efforts on the Critical Subproblems

The goal of all of these decomposition techniques is to divide a complex problem into simpler problems such that these simpler problems can be tackled in a focused way. Once problem decomposition is complete, the team chooses the subproblems that are most critical to the success of the product and that are most likely to benefit from novel or creative solutions. This approach involves a conscious decision to defer the solution of some of the subproblems. For example, the nailer team chose to focus on the subproblems of storing/accepting energy, converting the energy to translational energy, and applying the translational energy to the nail. The team felt confident that the nail handling and triggering issues could be solved after the energy storage and conversion issues were addressed. The team also deferred most of the user interaction issues of the tool. The team believed that the choice of a basic working principle for the tool would so constrain the eventual form of the tool that they had to begin with the core technology and then proceed to consider how to embody that technology in an attractive and user-friendly form. Teams can usually agree after a few minutes of discussion on which subproblems should be addressed first and which should be deferred for later consideration.

5.3 | STEP 2: SEARCH EXTERNALLY

External search is aimed at finding existing solutions to both the overall problem and to the subproblems identified during the problem clarification step. While external search is listed as the second step in the concept generation method, this sequential labeling is deceptive; external search occurs continually throughout the development process. Implementing an existing solution is usually quicker and cheaper than developing a new solution. Liberal use of existing solutions allows the team to focus its creative energy on the critical subproblems for which there are no satisfactory prior solutions. Furthermore, a conventional solution to one subproblem can frequently be combined with a novel solution to another subproblem to yield a superior overall design. For this reason external search includes detailed evaluation not only of directly competitive products but also of technologies used in products with related subfunctions.

The external search for solutions is essentially an information-gathering process. Available time and resources can be optimized by using an expand-and-focus strategy: first *expand* the scope of the search by broadly gathering information that might be related to the problem and then *focus* the scope of the search by exploring the promising directions in more detail. Too much of either approach will make the external search inefficient.

There are at least five good ways to gather information from external sources: lead user interviews, expert consultation, patent searches, literature searches, and competitive benchmarking.

Interview Lead Users

While identifying customer needs, the team may have sought out or encountered lead users. *Lead users* are those users of a product who experience needs months or years before the majority of the market and stand to benefit substantially from a product innovation (von Hippel, 1988). Frequently these lead users will have already invented solutions to meet their needs. This is particularly true among highly technical user communities, such as those in the medical or scientific fields. Lead users may be sought out in the market for which the team is developing the new product, or they may be found in markets for products implementing some of the subfunctions of the product.

In the hand-held nailer case, the nailer team consulted with the building contractors from the PBS television series *This Old House* in order to solicit new concepts. These lead users, who are exposed to tools from many manufacturers, made many interesting observations about the weaknesses in existing tools, but in this case did not provide many new product concepts.

Consult Experts

Experts with knowledge of one or more of the subproblems not only can provide solution concepts directly but also can redirect the search in a more fruitful area. Experts may include professionals at firms manufacturing related products, professional consultants, university faculty, and technical representatives of suppliers. These people can be found by calling universities, by calling companies, and by looking up authors of articles. While finding experts can be hard work, it is almost always less time consuming than re-creating existing knowledge.

Most experts are willing to talk on the telephone or meet in person for an hour or so without charge. In general, consultants will expect to be paid for time they spend on a problem beyond an initial meeting or telephone conversation. Suppliers are usually willing to provide several days of effort without direct compensation if they anticipate that someone will use their product as a component in a design. Of course, experts at directly competing firms are in most cases unwilling to provide proprietary information about their product designs. A good habit to develop is to always ask people consulted to suggest others who should be contacted. The best information often comes from pursuing these "second generation" leads.

The nailer design team consulted dozens of experts, including a rocket fuel specialist at Morton-Thiokol, electric motor researchers at MIT, and engineers from a vendor of gas springs. Most of this consultation was done on the telephone, although the engineers from the spring vendor made two trips to visit the team, at their company's expense.

Search Patents

Patents are a rich and readily available source of technical information containing detailed drawings and explanations of how many products work. The main disadvantage of patent searches is that concepts found in recent patents are protected (generally for 20 years from the date of the patent application), so there may be a royalty involved in using them. However, patents are also useful to see what concepts are already protected and must be avoided or licensed. Concepts contained in foreign patents without global coverage and in expired patents can be used without payment of royalties.

The formal indexing scheme for patents is difficult for novices to navigate. Fortunately, several databases contain the actual text of all patents. These text databases can be searched electronically by key words. Key word searches can be conducted efficiently with only modest practice and are remarkably effective in finding patents relevant to a particular product. Copies of U.S. patents including illustrations can be obtained for a nominal fee from the U.S. Patent and Trademark Office and from several suppliers. (See the web site www.ulrich-eppinger.net for a current list of on-line patent databases and suppliers of patent documents.)

A U.S. patent search in the area of nailers revealed several interesting concepts. One of the patents described a motor-driven double-flywheel nailer. One of the illustrations from this patent is shown in Figure 5.5. The design in this patent uses the accumulation of rotational kinetic energy in a flywheel, which is then suddenly converted into translational energy by a friction clutch. The energy is then delivered to the nail with a single impact of a drive pin.

Search Published Literature

Published literature includes journals; conference proceedings; trade magazines; government reports; market, consumer, and product information; and new product announcements. Literature searches are therefore very fertile sources of existing solutions.

Electronic searches are frequently the most efficient way to gather information from published literature. Searching the Internet is often a good first step, although the quality of the results can be hard to assess. More structured databases are available from online sources or on mass storage devices (e.g., CD-ROM). Many databases store only abstracts of articles and not the full text and diagrams. A follow-up search for an actual article is often needed for complete information. The two main difficulties in conducting good database searches are determining the key words and limiting the scope of the search. There is a trade-off between the need to use more key words for complete coverage and the need to restrict the number of matches to a manageable number.

Handbooks cataloging technical information can also be very useful references for external search. Examples of such engineering references are *Mark's Standard Handbook of Mechanical Engineering, Perry's Chemical Engineers' Handbook,* and *Mechanisms and Mechanical Devices Sourcebook.*

The nailer team found several useful articles related to the subproblems, including articles on energy storage describing flywheel and battery technologies. In a handbook they found an impact tool mechanism that provided a useful energy conversion concept.

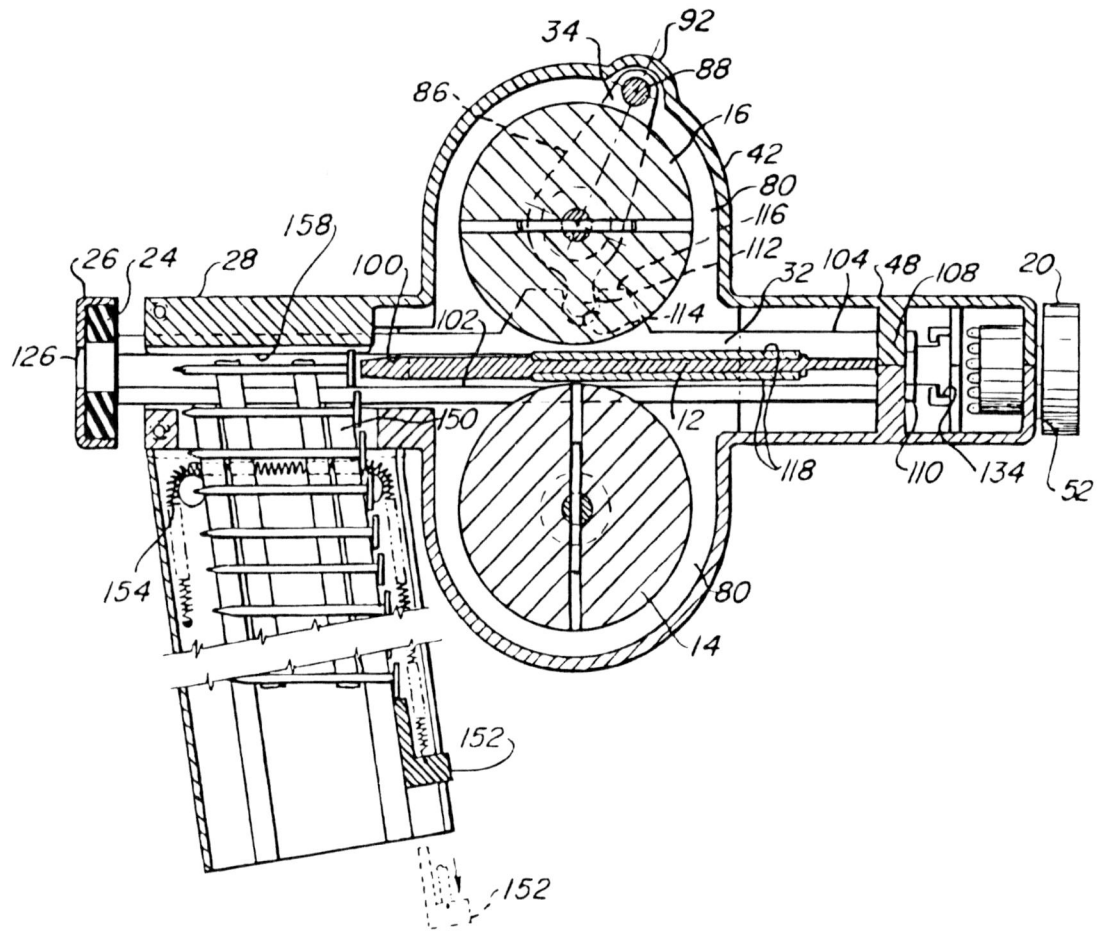

Figure 5.5
Concept from motor-driven double-flywheel nailer patent (U.S. Patent 4,042,036). The accompanying text describing the patent is nine pages long.

Benchmark Related Products

In the context of concept generation, *benchmarking* is the study of existing products with functionality similar to that of the product under development or to the subproblems on which the team is focused. Benchmarking can reveal existing concepts that have been implemented to solve a particular problem, as well as information on the strengths and weaknesses of the competition.

At this point the team will likely already be familiar with the competitive and closely related products. Products in other markets, but with related functionality, are more difficult to find. One of the most useful sources of this information is the *Thomas Register of American Manufacturers,* a directory of manufacturers of industrial products organized by product type. Often the hard-

est part of using the *Thomas Register* is finding out what related products are actually called and how they are cataloged. The *Thomas Register* can be accessed via the Internet.

For the nailer, the closely related products included a single-shot gunpowder-actuated tool for nailing into concrete, an electrical solenoid-actuated tacker, a pneumatic nailer for factory use, and a palm-held multiblow pneumatic nailer. The products with related functionality (in this case, energy storage and conversion) included air bags and the sodium azide propellant used as an energy source, chemical hand warmers for skiing, air rifles powered by carbon dioxide cartridges, and portable computers and their battery packs. The team obtained and disassembled most of these related products in order to discover the general concepts on which they were based, as well as other, more detailed information, including, for example, the names of the suppliers of specific components.

External search is an important method of gathering solution concepts. Skill in conducting external searches is therefore a valuable personal and organizational asset. This ability can be developed through careful observation of the world in order to develop a mental database of technologies and through the development of a network of professional contacts. Even with the aid of personal knowledge and contacts, external search remains "detective work" and is completed most effectively by those who are persistent and resourceful in pursuing leads and opportunities.

5.4 | STEP 3: SEARCH INTERNALLY

Internal search is the use of personal and team knowledge and creativity to generate solution concepts. The search is *internal* in that all of the ideas to emerge from this step are created from knowledge already in the possession of the team. This activity may be the most open-ended and creative of any in new-product development. We find it useful to think of internal search as a process of retrieving a potentially useful piece of information from one's memory and then adapting that information to the problem at hand. This process can be carried out by individuals working in isolation or by a group of people working together.

Four guidelines are useful for improving both individual and group internal search:

1. *Suspend judgment.* In most aspects of daily life, success depends on an ability to quickly evaluate a set of alternatives and take action. For example, none of us would be very productive if deciding what to wear in the morning or what to eat for breakfast involved an extensive period of generating alternatives before making a judgment. Because most decisions in our day-to-day lives have implications of only a few minutes or hours, we are accustomed to making decisions quickly and moving on. Concept generation for product development is fundamentally different. We have to live with the consequences of product concept decisions for years. As a result, suspending evaluation for the days or weeks required to generate a large set of alternatives is critical to success. The imperative to suspend judgment is frequently translated into the rule that during group concept generation sessions no criticism of concepts is allowed. A better approach is for individuals perceiving weak-

nesses in concepts to channel any judgmental tendencies into suggestions for improvements or alternative concepts.

2. *Generate a lot of ideas.* Most experts believe that the more ideas a team generates, the more likely the team is to explore fully the solution space. Striving for quantity lowers the expectations of quality for any particular idea and therefore may encourage people to share ideas they may otherwise view as not worth mentioning. Further, each idea acts as a stimulus for other ideas, so a large number of ideas has the potential to stimulate even more ideas.

3. *Welcome ideas that may seem infeasible.* Ideas which initially appear infeasible can often be improved, "debugged," or "repaired" by other members of the team. The more infeasible an idea, the more it stretches the boundaries of the solution space and encourages the team to think of the limits of possibility. Therefore, infeasible ideas are quite valuable and their expression should be encouraged.

4. *Use graphical and physical media.* Reasoning about physical and geometric information with words is difficult. Text and verbal language are inherently inefficient vehicles for describing physical entities. Whether working as a group or as an individual, abundant sketching surfaces should be available. Foam, clay, cardboard, and other three-dimensional media may also be appropriate aids for problems requiring a deep understanding of form and spatial relationships.

Both Individual and Group Sessions Can Be Useful

Formal studies of group and individual problem solving suggest that a set of people working alone for a period of time will generate more and better concepts than the same people working together for the same time period (McGrath, 1984). This finding is contrary to the actual practices of the many firms that perform most of their concept generation activities in group sessions. Our observations confirm the formal studies, and we believe that team members should spend at least some of their concept generation time working alone. We also believe that group sessions are critical for building consensus, communicating information, and refining concepts. In an ideal setting, each individual on the team would spend several hours working alone and then the group would get together to discuss and improve the concepts generated by individuals.

However, we also know that there is a practical reason for holding group concept generation sessions: it is one way to guarantee that the individuals in the group will devote a certain amount of time to the task. Especially in very intense and demanding work environments, without scheduling a meeting, few people will allocate several hours for concentrated individual effort on generating new concepts. The phone rings, people interrupt, urgent problems demand attention. In certain environments, scheduled group sessions may be the only way to guarantee that enough attention is paid to the concept generation activity.

The nailer team used both individual effort and group sessions for internal search. For example, during one particular week each member was assigned one or two subproblems and was expected to develop at least 10 solution concepts. This divided the concept generation work among all members. The group then met to discuss and expand on the individually generated concepts. The more promising concepts were investigated further.

Hints for Generating Solution Concepts

Experienced individuals and teams can usually just sit down and begin generating good concepts for a subproblem. Often these people have developed a set of techniques they use to stimulate their thinking, and these techniques have become a natural part of their problem-solving process. Novice product development professionals may be aided by a set of hints that stimulate new ideas or encourage relationships among ideas. VanGundy (1988), von Oech (1983), and McKim (1980) give dozens of helpful suggestions. Here are some hints we have found to be helpful:

- *Make analogies.* Experienced designers always ask themselves what other devices solve a related problem. Frequently they will ask themselves if there is a natural or biological analogy to the problem. They will think about whether their problem exists at a much larger or smaller dimensional scale than that which they are considering. They will ask what devices do something similar in an unrelated area of application. The nailer team, when posing these questions, realized that construction pile drivers are similar to nailers in some respects. In following up on this idea, they developed the concept of a multiblow tool.
- *Wish and wonder.* Beginning a thought or comment with "I wish we could . . ." or "I wonder what would happen if . . ." helps to stimulate oneself or the group to consider new possibilities. These questions cause reflection on the boundaries of the problem. For example, a member of the nailer team, when confronted with the required length of a rail gun (an electromagnetic device for accelerating a projectile) for driving a nail, said, "I wish the tool could be 1 meter long." Discussion of this comment led to the idea that perhaps a long tool could be used like a cane for nailing decking, allowing users to remain on their feet.
- *Use related stimuli.* Most individuals can think of a new idea when presented with a new stimulus. Related stimuli are those stimuli generated in the context of the problem at hand. For example, one way to use related stimuli is for each individual in a group session to generate a list of ideas (working alone) and then pass the list to his or her neighbor. Upon reflection on someone else's ideas, most people are able to generate new ideas. Other related stimuli include customer needs statements and photographs of the use environment of the product.
- *Use unrelated stimuli.* Occasionally, random or unrelated stimuli can be effective in encouraging new ideas. An example of such a technique is to choose, at random, one of a collection of photographs of objects, and then to think of some way that the randomly generated object might relate to the problem at hand. In a variant of this idea, individuals can be sent out on the streets with an instant camera to capture random images for subsequent use in stimulating new ideas. (This may also serve as a good change of pace for a tired group.)
- *Set quantitative goals.* Generating new ideas can be exhausting. Near the end of a session, individuals and groups may find quantitative goals useful as a motivating force. The nailer team frequently issued individual concept generation assignments with quantitative targets of 10 to 20 concepts.
- *Use the gallery method.* The *gallery method* is a way to display a large number of concepts simultaneously for discussion. Sketches, usually one concept to a sheet, are taped or pinned to the walls of the meeting room. Team members circulate and look at each concept. The creator of the concept may offer explanation, and the group subsequently makes suggestions for improving the concept or spontaneously generates related concepts. This method is a good way to merge individual and group efforts.

In the 1990s, a Russian problem-solving methodology called TRIZ (a Russian acronym for *theory of inventive problem solving*) began to be disseminated in the United States. The methodology is primarily useful in identifying physical working principles to solve technical problems. The key idea underlying TRIZ is to identify a contradiction that is implicit in a problem. For example, a contra-

Solutions to Subproblem of Storing or Accepting Energy

- Self-regulating chemical reaction emitting high-pressure gas
- Carbide (as for lanterns)
- Combusting sawdust from job site
- Gun powder
- Sodium azide (air bag explosive)
- Fuel-air combustion (butane, propane, acetylene, etc.)
- Compressed air (in tank or from compressor)
- Carbon dioxide in tank
- Electric wall outlet and cord
- High-pressure oil line (hydraulics)
- Flywheel with charging (spin-up)
- Battery pack or tool, belt, or floor
- Fuel cell
- Human power: arms or legs
- Methane from decomposing organic materials
- "Burning" like that of chemical hand warmers
- Nuclear reactions
- Cold fusion
- Solar electric cells
- Solar-steam conversion
- Steam supply line
- Wind
- Geothermal

Solutions to Subproblem of Applying Translational Energy to Nail

Single Impact

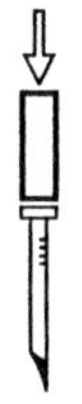

Multiple Impacts
(tens or hundreds)

Multiple Impacts
(hundreds or thousands)

Push

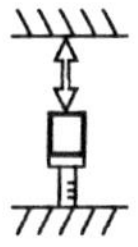

Twist-Push

Figure 5.6
Some of the solutions to the subproblems of (1)storing or accepting energy and (2)delivering translational energy to a nail.

diction in the nailer problem might be that increasing power (a desirable characteristic) would also tend to increase weight (an undesirable characteristic). One of the TRIZ tools is a matrix of 39 by 39 characteristics with each cell corresponding to a particular conflict between two characteristics. In each cell of the matrix, up to four physical principles are suggested as ways of resolving the corresponding conflict. There are 40 basic principles, including, for example, the *periodic action* principle (i.e., replace a continuous action with a periodic action, like an impulse). Using TRIZ, the nailer team might have arrived at the concept of using repeated smaller impacts to drive the nail. The idea of identifying a conflict in the design problem and then thinking about ways to resolve the con-

flict appears to be a very useful problem-solving heuristic. This approach can be useful in generating concepts even without adopting the entire TRIZ methodology.

Figure 5.6 shows some of the solutions the nailer team generated for the subproblems of (1)storing or accepting energy and (2)delivering translational energy to a nail.

5.5 | STEP 4: EXPLORE SYSTEMATICALLY

As a result of the external and internal search activities, the team will have collected tens or hundreds of concept *fragments*—solutions to the subproblems. Systematic exploration is aimed at navigating the space of possibilities by organizing and synthesizing these solution fragments. The nailer team focused on the energy storage, conversion, and delivery subproblems and had generated dozens of concept fragments for each subproblem. One approach to organizing and synthesizing these fragments would be to consider all of the possible combinations of the fragments associated with each subproblem; however, a little arithmetic reveals the impossibility of this approach. Given the three subproblems on which the team focused and an average of 15 fragments for each subproblem, the team would have to consider 3,375 combinations of fragments ($15 \times 15 \times 15$). This would be a daunting task for even the most enthusiastic team. Furthermore, the team would quickly discover that many of the combinations do not even make sense. Fortunately, there are two specific tools for managing this complexity and organizing the thinking of the team: the *concept classification tree* and the *concept combination table*. The classification tree helps the team divide the possible solutions into independent categories. The combination table guides the team in selectively considering combinations of fragments.

Concept Classification Tree

The concept classification tree is used to divide the entire space of possible solutions into several distinct classes which will facilitate comparison and pruning. An example of a tree for the nailer example is shown in Figure 5.7. The branches of this tree correspond to different energy sources.

The classification tree provides at least four important benefits:

1. *Pruning of less promising branches:* If by studying the classification tree the team is able to identify a solution approach that does not appear to have much merit, then this approach can be pruned and the team can focus its attention on the more promising branches of the tree. Pruning a branch of the tree requires some evaluation and judgment and should therefore be done carefully, but the reality of product development is that there are limited resources and that focusing the available resources on the most promising directions is an important success factor. For the nailer team, the nuclear energy source was pruned from consideration. Although the team had identified some very intriguing nuclear devices for use in powering artificial hearts, they felt that these devices would not be economically practical for at least a decade and

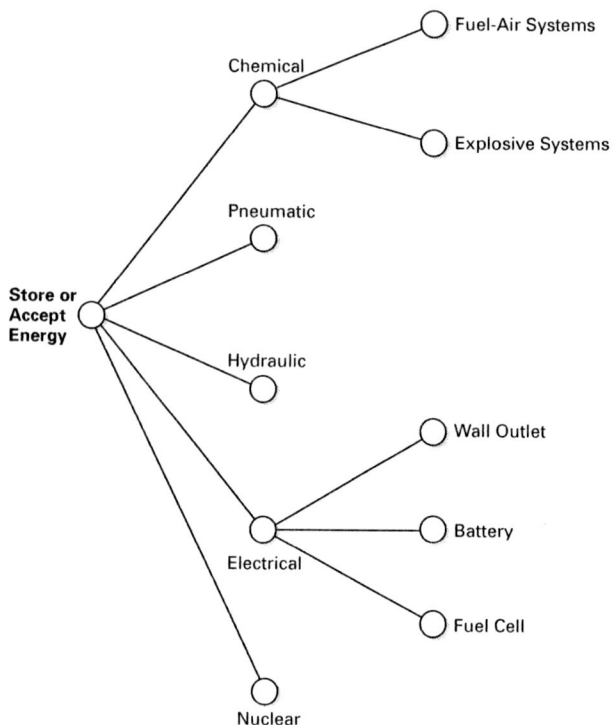

Figure 5.7
A classification tree for the nailer energy source concept fragments.

would probably be hampered by regulatory requirements indefinitely.

2. *Identification of independent approaches to the problem:* Each branch of the tree can be considered a different approach to solving the overall problem. Some of these approaches may be almost completely independent of each other. In these cases, the team can cleanly divide its efforts among two or more individuals or task forces. When two approaches both look promising, this division of effort can reduce the complexity of the concept generation activities. It also may engender some healthy competition among the approaches under consideration. The nailer team found that both the chemical/explosive branch and the electrical branch appeared quite promising. They assigned these two approaches to two different subteams and pursued them independently for several weeks.

3. *Exposure of inappropriate emphasis on certain branches:* Once the tree is constructed, the team is able to reflect quickly on whether the effort applied to each branch has been appropriately allocated.

The nailer team recognized that they had applied very little effort to thinking about hydraulic energy sources and conversion technologies. This recognition guided them to focus on this branch of the tree for a few days.

4. *Refinement of the problem decomposition for a particular branch:* Sometimes a problem decomposition can be usefully tailored to a particular approach to the problem. Consider the branch of the tree corresponding to the electrical energy source. Based on additional investigation of the nailing process, the team determined that the instantaneous power delivered during the nailing process was about 10,000 watts for a few milliseconds and so exceeds the power which is available from a wall outlet, a battery, or a fuel cell (of reasonable size, cost, and mass). They concluded, therefore, that energy must be accumulated over a substantial period of the nailing cycle (say 100 milliseconds) and then suddenly released to supply the required instantaneous power to drive the nail. This quick analysis led the team to add a subfunction ("accumulate translational energy") to their function diagram (see Figure 5.8). They chose to add the subfunction after the conversion of electrical energy to mechanical energy, but briefly considered the possibility of accumulating the energy in the electrical domain with a capacitor. This kind of refinement of the function diagram is quite common as the team makes more assumptions about the approach and as more information is gathered.

The classification tree in Figure 5.7 shows the alternative solutions to the energy source subproblem. However, there are other possible trees. The team might have chosen to use a tree classifying the alternative solutions to the energy delivery subproblem, showing branches for single impact, multiple impact, or pushing. Trees can be constructed with branches corresponding to the solution fragments of any of the subproblems, but certain classifica-

tions are more useful. In general, a subproblem whose solution highly constrains the possible solutions to the remaining subproblems is a good candidate for a classification tree. For example, the choice of energy source (electrical, nuclear, pneumatic, etc.) constrains whether a motor or a piston-cylinder can be used to convert the energy to translational energy. In contrast, the choice of energy delivery mechanism (single impact, multiple impact, etc.) does not greatly constrain the solutions to the other subproblems. Reflection on which subproblem is likely to most highly constrain the solutions to the remaining subproblems will usually lead to one or two clear ways to construct the classification tree.

Concept Combination Table

The concept combination table provides a way to consider combinations of solution fragments systematically. Figure 5.9 shows an example of a combination table that the nailer team used to consider the combinations of fragments for the electrical branch of the classification tree. The columns in the table correspond to the subproblems identified in Figure 5.8. The entries in each column correspond to the solution fragments for each of these subproblems derived from external and internal search. For example, the subproblem of converting electrical energy to translational energy is the heading for the first column. The entries in this column are a rotary motor with a transmission, a linear motor, a solenoid, and a rail gun.

Potential solutions to the overall problem are formed by combining one fragment from each column. For the nailer example, there are 24 possible combinations ($4 \times 2 \times 3$). Choosing a combination of fragments does not lead spontaneously to a solution to the overall problem. The combination of fragments must usually be developed and refined before an integrated solution emerges. This development may not even be possible or may lead to more than one solution, but at a minimum it involves additional creative thought. In some ways, the combination table is sim-

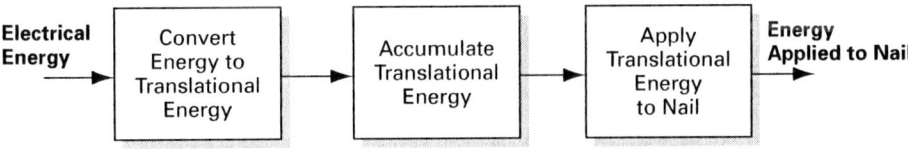

Figure 5.8
A new problem decomposition assuming an electrical energy source and the accumulation of energy in the mechanical domain.

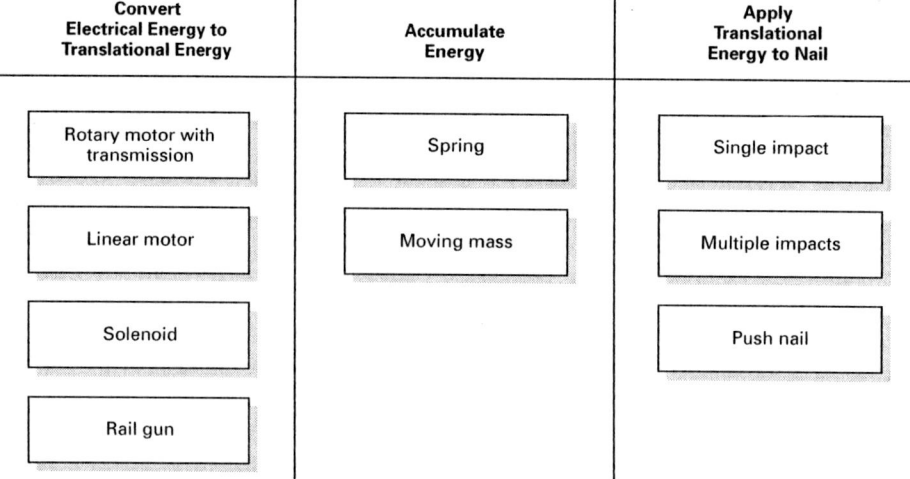

Figure 5.9

Concept combination table for the hand-held nailer.

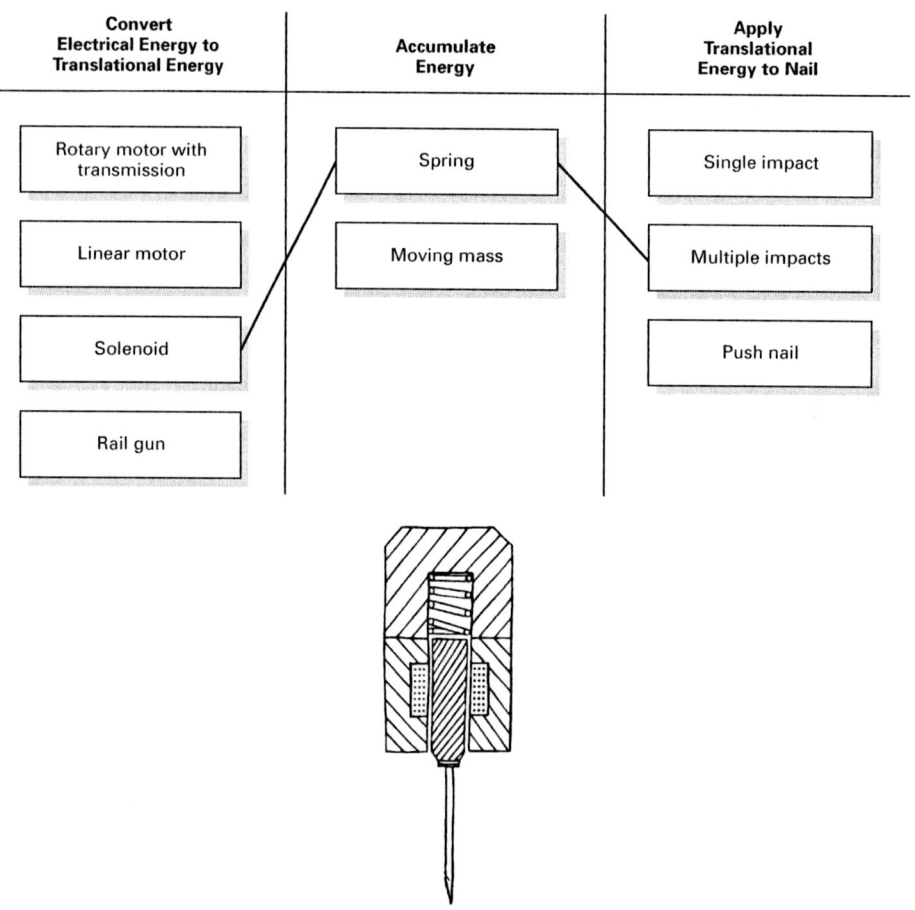

Figure 5.10

In this solution concept, a solenoid compresses a spring and then releases it repeatedly in order to drive the nail with multiple impacts.

ply a way to make forced associations among fragments in order to stimulate further creative thinking; in no way does the mere act of selecting a combination yield a complete solution.

Figure 5.10 shows a sketch of a concept arising from the combination of the fragments "solenoid," "spring," and "multiple impacts." Figure 5.11 shows some sketches of concepts arising from the combination of the fragments "rotary motor with transmission," "spring," and "single impact." Figure 5.12 shows a sketch of a concept arising from the combination of "rotary motor with transmission," "spring," and "multiple impacts." Figure 5.13 shows some sketches of concepts arising from the combination of "linear motor," "moving mass," "and single impact."

Two guidelines make the concept combination process easier. First, if a fragment can be eliminated as being infeasible before combining it with other fragments, then the number of combinations the team needs to consider is dramatically reduced. For example, if the team could determine that the rail gun would not be feasible under any condition, they could reduce the number of combinations from 24 to 18. Second, the concept combination table should be concentrated on the subproblems that are coupled. Coupled subproblems are those whose solutions can be evaluated only in combination with the solutions to other subproblems. For example, the choice of the specific electrical energy source to be used (e.g., battery versus wall outlet), although extremely critical, is somewhat in-

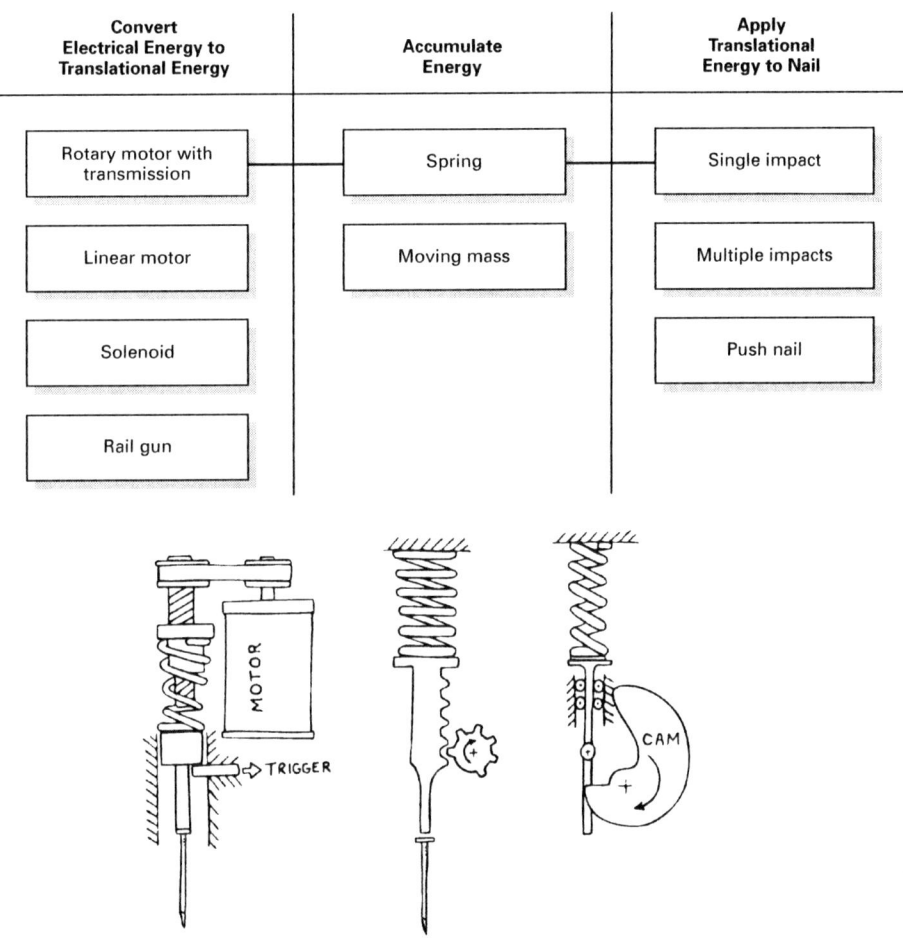

Figure 5.11
Multiple solutions arising from the combination of a motor with transmission, a spring, and single impact. The motor winds a spring, accumulating potential energy which is then delivered to the nail in a single blow.

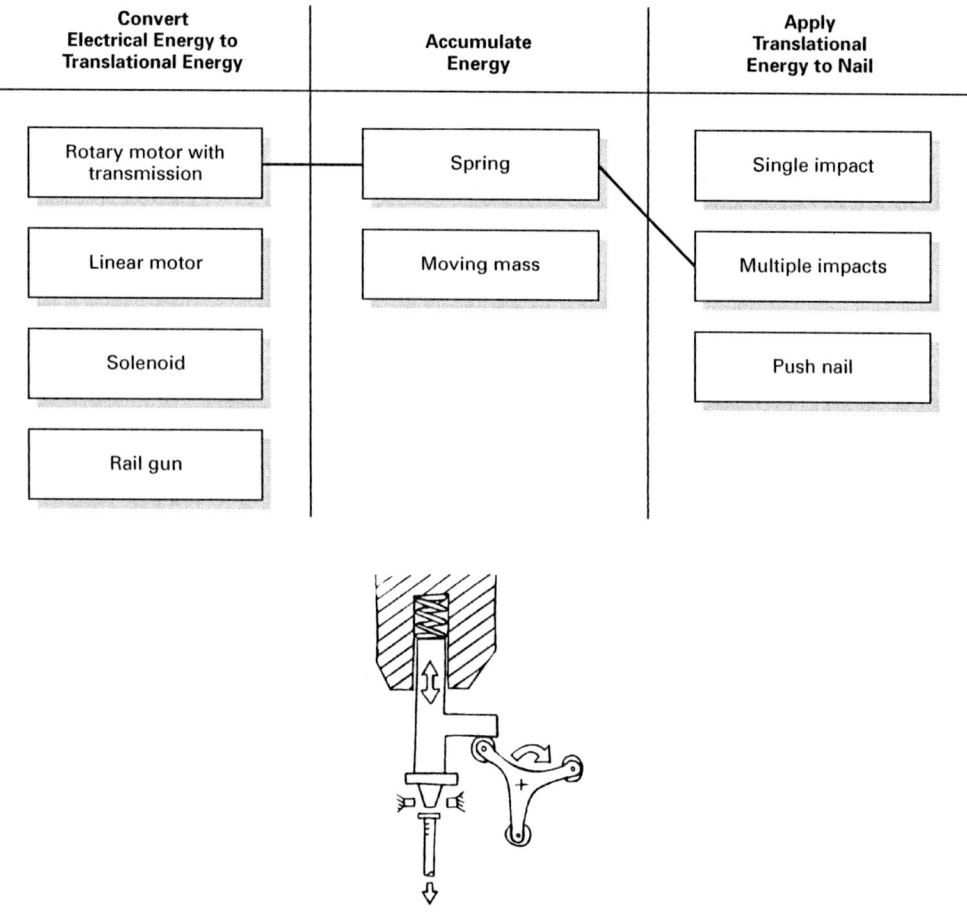

Convert Electrical Energy to Translational Energy	Accumulate Energy	Apply Translational Energy to Nail
Rotary motor with transmission	Spring	Single impact
Linear motor	Moving mass	Multiple impacts
Solenoid		Push nail
Rail gun		

Figure 5.12

Solution from the combination of a motor with transmission, a spring, and multiple impacts. The motor repeatedly winds and releases the spring, storing and delivering energy over several blows.

dependent of the choice of energy conversion (e.g., motor versus solenoid). Therefore, the concept combination table does not need to contain a column for the different types of electrical energy sources. This reduces the number of combinations the team must consider. As a practical matter, concept combination tables lose their usefulness when the number of columns exceeds three or four.

Managing the Exploration Process

The classification tree and combination tables are tools that a team can use somewhat flexibly. They are simple ways to organize thinking and guide the creative energies of the team. Rarely do teams generate only one classification tree and one concept combination table. More typically the team will create several alternative classification trees and several concept combination tables. Interspersed with this exploratory activity may be a refining of the original problem decomposition or the pursuit of additional internal or external search. The exploration step of concept generation usually acts more as a guide for further creative thinking than as the final step in the process.

Recall that at the beginning of the process the team chooses a few subproblems on which to focus attention. Eventually the team must return to address all of the subproblems. This usually occurs after the team has narrowed the range of alternatives for the critical subproblems. The nailer team narrowed its alternatives to a few chemical and a few electric concepts and then refined them by working out the user interface, industrial design, and configuration issues. One of the resulting concept descriptions is shown in Figure 5.14.

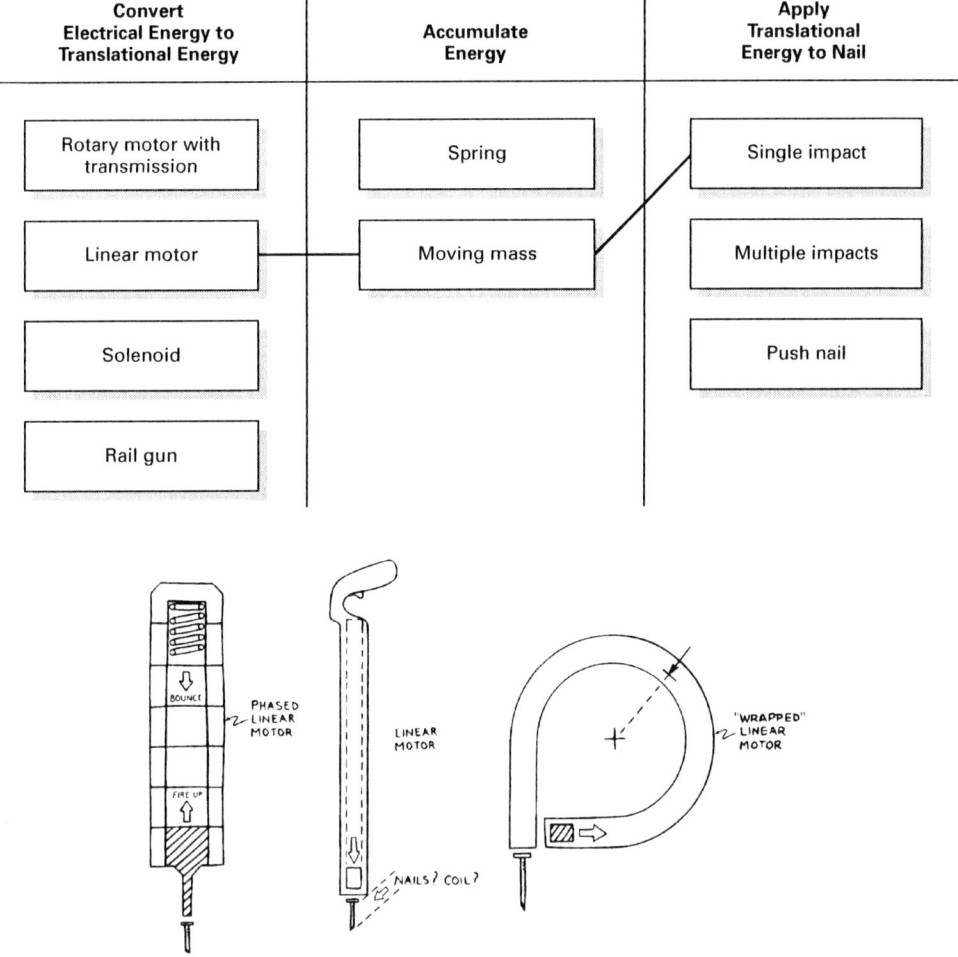

Figure 5.13
Solutions from the combination of a linear motor, a moving mass, and single impact. A linear motor accelerates a massive hammer, accumulating kinetic energy which is delivered to the nail in a single blow.

5.6 | STEP 5: REFLECT ON THE RESULTS AND THE PROCESS

Although the reflection step is placed here at the end for convenience in presentation, reflection should in fact be performed throughout the whole process. Questions to ask include:

- Is the team developing confidence that the solution space has been fully explored?
- Are there alternative function diagrams?
- Are there alternative ways to decompose the problem?
- Have external sources been thoroughly pursued?
- Have ideas from everyone been accepted and integrated in the process?

The nailer team members discussed whether they had focused too much attention on the energy storage and conversion issues in the tool while ignoring the user interface and overall configuration. They decided that the energy issues remained at the core of the problem and that their decision to focus on these issues first was justified. They also wondered if they had pursued too many branches of the classification tree. Initially they had pursued electrical, chemical, and pneumatic concepts before ultimately settling on an electric concept. In hindsight, the chemical approach had some obvious safety and customer perception shortcomings (they were exploring the use of explosives as an energy source). They decided that although they liked some aspects of the chemical solution, they should have eliminated it from consideration earlier in the

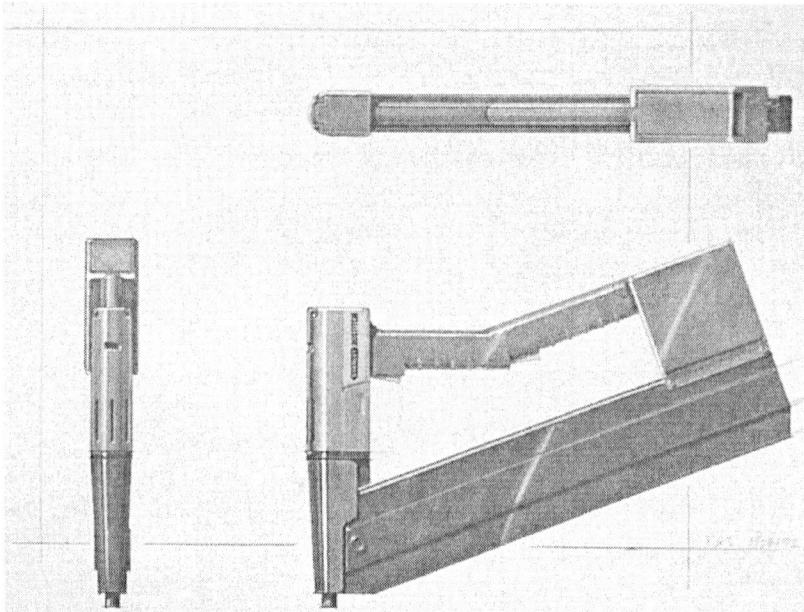

Figure 5.14
One of several refined solution concepts.
(Courtesy of Product Genesis Inc.)

process, allowing more time to pursue some of the more promising branches in greater detail.

5.7 | SUMMARY

A product concept is an approximate description of the technology, working principles, and form of the product. The degree to which a product satisfies customers and can be successfully commercialized depends to a large measure on the quality of the underlying concept.

- The concept generation process begins with a set of customer needs and target specifications and results in a set of product concepts from which the team will make a final selection.
- In most cases, an effective development team will generate hundreds of concepts, of which 5 to 20 will merit serious consideration during the subsequent concept selection activity.
- The concept generation method presented in this chapter consists of five steps:
  1. *Clarify the problem.* Understand the problem and decompose it into simpler subproblems.
  2. *Search externally.* Gather information from lead users, experts, patents, published literature, and related products.
  3. *Search internally.* Use individual and group methods to retrieve and adapt the knowledge of the team.
  4. *Explore systematically.* Use classification trees and combination tables to organize the thinking of the team and to synthesize solution fragments.
  5. *Reflect on the solutions and the process.* Identify opportunities for improvement in subsequent iterations or future projects.
- Although concept generation is an inherently creative process, teams can benefit from using a structured method. Such an approach allows full exploration of the design space and reduces the chance of oversight in the types of solution concepts considered. It also acts as a map for those team members who are less experienced in design problem solving.
- Despite the linear presentation of the concept generation process in this chapter, the team will likely return to each step of the process several times. Iteration is particularly common when the team is developing a radically new product.
- Professionals who are good at concept generation seem to always be in great demand as team members. Contrary to popular opinion, we believe concept generation is a skill that can be learned and developed.

5.8 | REFERENCES AND BIBLIOGRAPHY

Many current resources are available on the Internet via

www.ulrich-eppinger.net

Pahl and Beitz were the driving force behind structured design methods in Germany. We adapt many of their ideas for functional decomposition.

Pahl, Gerhard, and Wolfgang Beitz, *Engineering Design,* second edition, K. Wallace, editor, Springer-Verlag, New York, 1996.

Hubka and Eder have written in a detailed way about systematic concept generation for technical products.

Hubka, Vladimir, and W. Ernst Eder, *Theory of Technical Systems: A Total Concept Theory for Engineering Design,* Springer-Verlag, New York, 1988.

Von Hippel reports on his empirical research on the sources of new product concepts. His central argument is that lead users are the innovators in many markets.

von Hippel, Eric, *The Sources of Innovation,* Oxford University Press, New York, 1988.

VanGundy presents dozens of methods for problem solving, many of which are directly applicable to product concept generation.

VanGundy, Arthur B., Jr., *Techniques of Structured Problem Solving,* second edition, Van Nostrand Reinhold, New York, 1988.

Von Oech provides dozens of good ideas for improving individual and group creative performance.

von Oech, Roger, *A Whack on the Side of the Head: How You Can be More Creative,* revised edition, Warner Books, New York, 1998.

McKim presents a holistic approach to developing creative thinking skills in individuals and groups.

McKim, Robert H., *Experiences in Visual Thinking,* second edition, Brooks/Cole Publishing, Monterey, CA, 1980.

Interesting research on a set of standard "templates" for identifying novel product concepts has been done by Goldenberg and his colleagues.

Goldenberg, Jacob, David Mazursky, and Sorin Solomon, "Toward Identifying the Inventive Templates of New Products: A Channeled Ideation Approach," *Journal of Marketing Research,* Vol. 36, No. 2, 1999.

The following are two of the better English-language publications on TRIZ.

Altshuller, Genrich, *40 Principles: TRIZ Keys to Technical Innovation,* Technical Innovation Center, Worcester, MA, 1998.

Terninko, John, Alla Zusman, and Boris Zlotin, *Systematic Innovation: An Introduction to TRIZ,* St. Lucie Press, Boca Raton, 1998.

McGrath presents studies comparing the relative performance of groups and individuals in generating new ideas.

McGrath, Joseph, E., *Groups: Interaction and Performance,* Prentice Hall, Englewood Cliffs, NJ, 1984.

Engineering handbooks are handy sources of information on standard technical solutions. Three good handbooks are:

Avallone, Eugene A., and Theodore Baumeister III (eds.), *Mark's Standard Handbook of Mechanical Engineering,* 10th edition, McGraw-Hill, New York, 1996.

Perry, Robert H., Don W. Green, James O. Maloney (eds.), *Perry's Chemical Engineers' Handbook,* seventh edition, McGraw-Hill, New York, 1997.

Chironis, Nicholas P., *Mechanisms and Mechanical Devices Sourcebook,* second edition, McGraw-Hill, New York, 1996.

Exercises

1. Decompose the problem of designing a new barbecue grill. Try a functional decomposition as well as a decomposition based on the user interactions with the product.
2. Generate 20 concepts for the subproblem "prevent fraying of end of rope" as part of a system for cutting lengths of nylon rope from a spool.
3. Prepare an external-search plan for the problem of permanently applying serial numbers to plastic products.

Thought Questions

1. What are the prospects for computer support for concept generation activities? Can you think of any computer tools that would be especially helpful in this process?
2. What would be the relative advantages and disadvantages of involving actual customers in the concept generation process?
3. For what types of products would the initial focus of the concept generation activity be on the form and user interface of the product and not on the core technology? Describe specific examples.

4. Could you apply the five-step method to an everyday problem like choosing the food for a picnic?
5. Consider the task of generating new concepts for the problem of dealing with leaves on a lawn. How would a plastic-bag manufacturer's assumptions and problem decomposition differ from those of a manufacturer of lawn tools and equipment and from those of a company responsible for maintaining golf courses around the world? Should the context of the firm dictate the way concept generation is approached?

Chapter

Concept Selection 6

A medical supply company retained a design firm to develop a reusable syringe with precise dosage control for outpatient use. One of the products sold by a competitor is shown in Figure 6.1. To focus the development effort, the medical supply company identified two major problems with its current product: cost (the existing model was made of stainless steel) and accuracy of dose metering. The company also requested that the product be tailored to the physical capabilities of the elderly, an important segment of the target market. To summarize the needs of its client and of the intended end users, the team established seven criteria on which the choice of a product concept would be based:

- Ease of handling.
- Ease of use.
- Readability of dose settings.

- Dose metering accuracy.
- Durability.
- Ease of manufacture.
- Portability.

The team described the concepts under consideration with the sketches shown in Figure 6.2. Although each concept nominally satisfied the key customer needs, the team was faced with choosing the best concept for further design, refinement, and production. The need to select one syringe concept from many raises several questions:

- How can the team choose the best concept, given that the designs are still quite abstract?
- How can a decision be made that is embraced by the whole team?

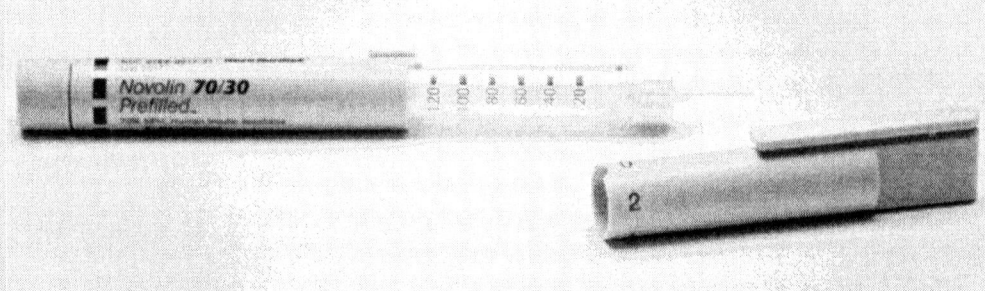

Figure 6.1
One of the existing outpatient syringes.
(Courtesy of Novo Nordisk Pharmaceuticals Inc.)

- How can desirable attributes of otherwise weak concepts be identified and used?
- How can the decision-making process be documented?

This chapter uses the syringe example to present a concept selection methodology addressing these and other issues.

6.1 | CONCEPT SELECTION IS AN INTEGRAL PART OF THE PRODUCT DEVELOPMENT PROCESS

Early in the development process the product development team identifies a set of customer needs. By using a

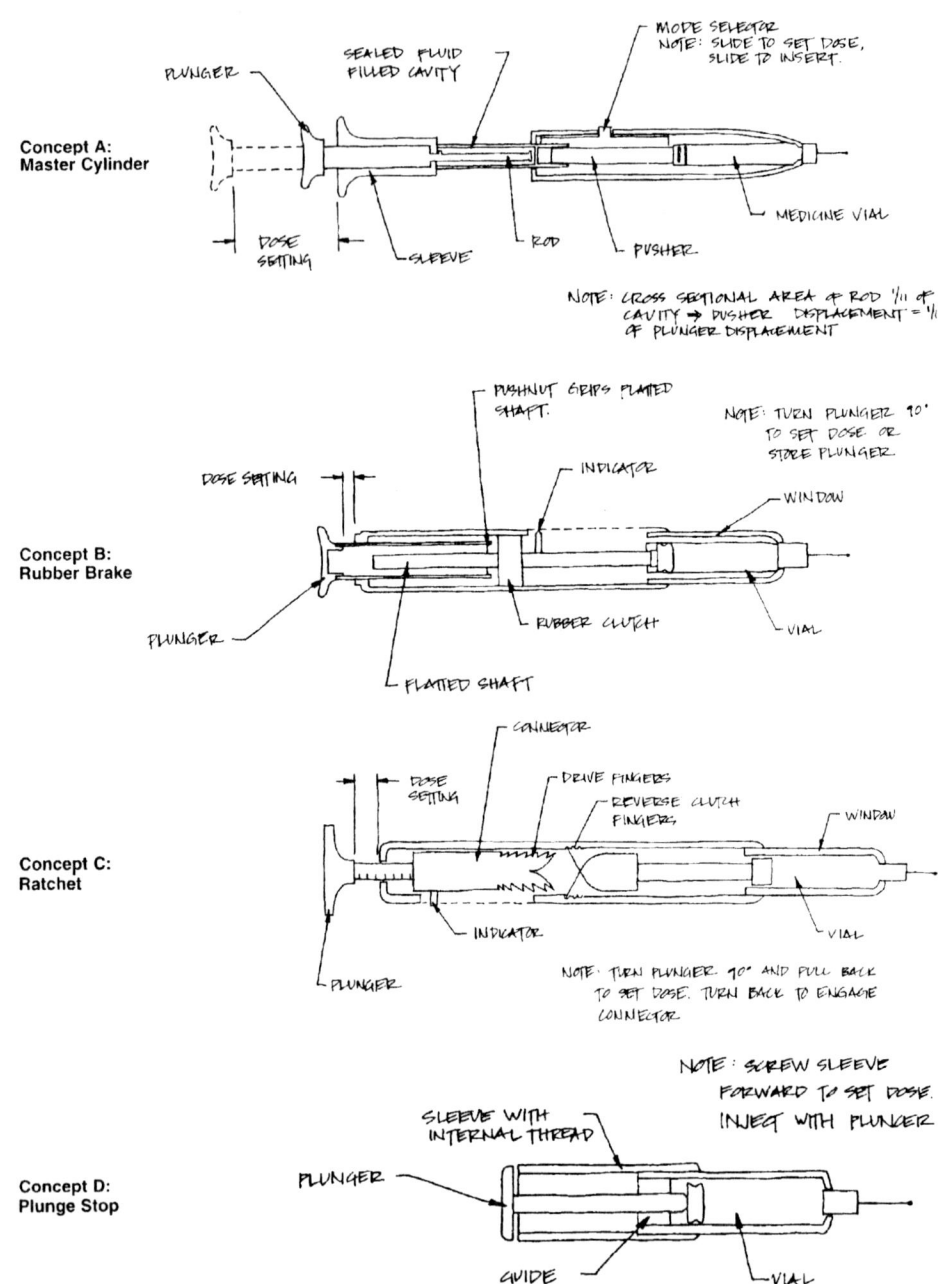

Figure 6.2

Seven concepts for the outpatient syringe. The product development team generated the seven sketches to describe the basic concepts under consideration.

variety of methods, the team then generates alternative solution concepts in response to these needs. (See Chapter 4, Identifying Customer Needs, and Chapter 6, Concept Generation, for more detail on these activities.) *Concept selection* is the process of evaluating concepts with respect to customer needs and other criteria, comparing the relative strengths and weaknesses of the concepts, and selecting one or more concepts for further investigation, testing, or development. Figure 6.3 illustrates how the concept selection activity is related to the other activities that make up the concept development phase of the product development process. Although this chapter focuses on the selection of an overall product concept at the beginning of the development process, the method we present is also useful later in the development process when the team must select subsystem concepts, components, and production processes.

While many stages of the development process benefit from unbounded creativity and divergent thinking, concept selection is the process of narrowing the set of con-

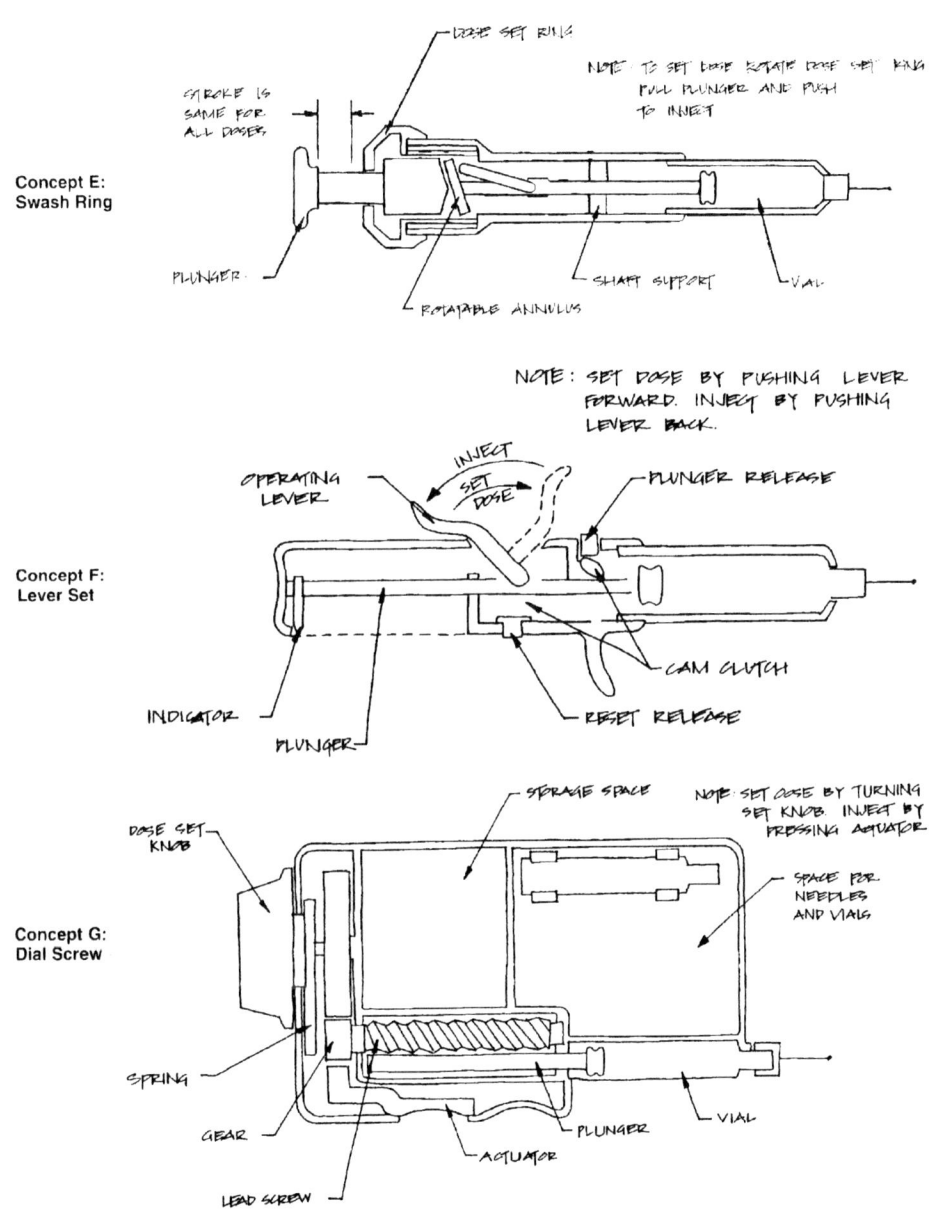

Figure 6.2 *Continued*

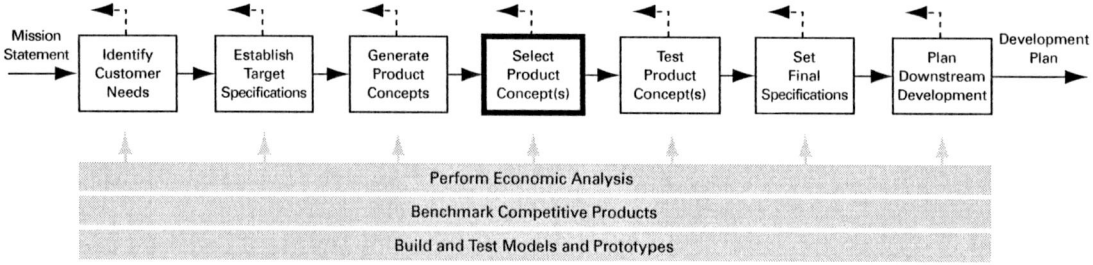

Figure 6.3
Concept selection is part of the overall concept development phase.

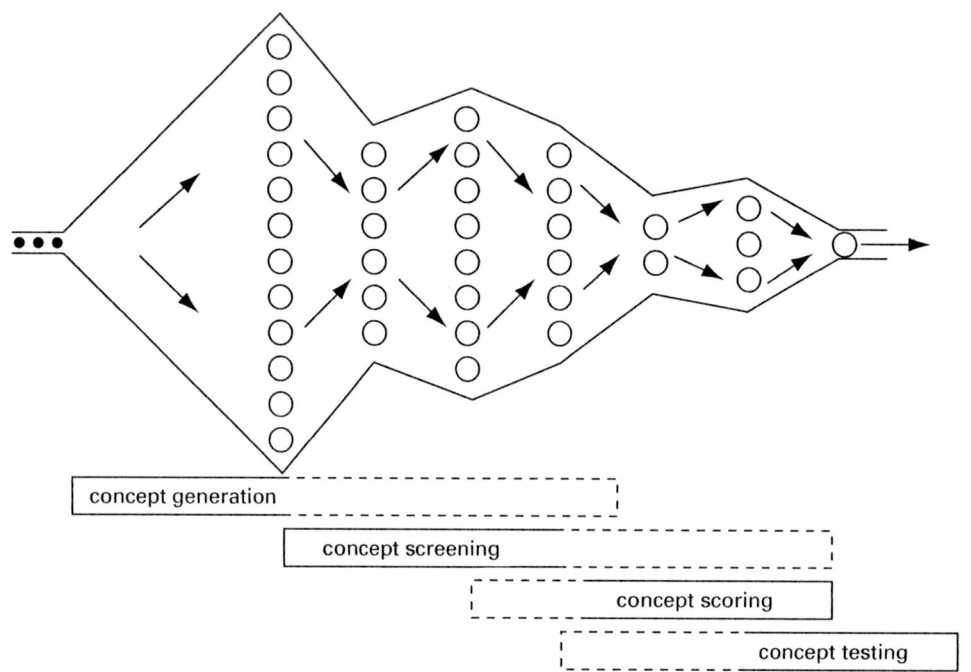

Figure 6.4
Concept selection is an iterative process closely related to concept generation and testing. The concept screening and scoring methods help the team refine and improve the concepts, leading to one or more promising concepts upon which further testing and development activities will be focused.

cept alternatives under consideration. Although concept selection is a convergent process, it is frequently iterative and may not produce a dominant concept immediately. A large set of concepts is initially winnowed down to a smaller set, but these concepts may subsequently be combined and improved to temporarily enlarge the set of concepts under consideration. Through several iterations a dominant concept is finally chosen. Figure 6.4 illustrates the successive narrowing and temporary widening of the set of options under consideration during the concept selection activity.

6.2 | ALL TEAMS USE SOME METHOD FOR CHOOSING A CONCEPT

Whether or not the concept selection process is explicit, all teams use some method to choose among concepts. (Even those teams generating only one concept are using a method: choosing the first concept they think of.) The methods vary in their effectiveness and include the following:

- **External decision:** Concepts are turned over to the customer, client, or some other external entity for selection.

- *Product champion:* An influential member of the product development team chooses a concept based on personal preference.
- *Intuition:* The concept is chosen by its feel. Explicit criteria or trade-offs are not used. The concept just *seems* better.
- *Multivoting:* Each member of the team votes for several concepts. The concept with the most votes is selected.
- *Pros and cons:* The team lists the strengths and weaknesses of each concept and makes a choice based upon group opinion.
- *Prototype and test:* The organization builds and tests prototypes of each concept, making a selection based upon test data.
- *Decision matrices:* The team rates each concept against prespecified selection criteria, which may be weighted.

The concept selection method in this chapter is built around the use of decision matrices for evaluating each concept with respect to a set of selection criteria.

6.3 | A STRUCTURED METHOD OFFERS SEVERAL BENEFITS

All of the early phases of product development are extremely influential on eventual product success. Certainly the response of the marketplace to a product depends critically on the product concept, but many practitioners and researchers also believe that the choice of a product concept dramatically constrains the eventual manufacturing cost of the product. A structured concept selection process helps to maintain objectivity throughout the concept phase of the development process and guides the product development team through a critical, difficult, and sometimes emotional process. Specifically, a structured concept selection method offers the following potential benefits:

- *A customer-focused product:* Because concepts are explicitly evaluated against customer-oriented criteria, the selected concept is likely to be focused on the customer.
- *A competitive design:* By benchmarking concepts with respect to existing designs, designers push the design to match or exceed their competitors' performance along key dimensions.
- *Better product-process coordination:* Explicit evaluation of the product with respect to manufacturing criteria improves the product's manufacturability and helps to match the product with the process capabilities of the firm.

- *Reduced time to product introduction:* A structured method becomes a common language among design engineers, manufacturing engineers, industrial designers, marketers, and project managers, resulting in decreased ambiguity, faster communication, and fewer false starts.
- *Effective group decision making:* Within the development team, organizational philosophy and guidelines, willingness of members to participate, and team member experience may constrain the concept selection process. A structured method encourages decision making based on objective criteria and minimizes the likelihood that arbitrary or personal factors influence the product concept.
- *Documentation of the decision process:* A structured method results in a readily understood archive of the rationale behind concept decisions. This record is useful for assimilating new team members and for quickly assessing the impact of changes in the customer needs or in the available alternatives.

6.4 | OVERVIEW OF METHODOLOGY

We present a two-stage concept selection methodology, although the first stage may suffice for simple design decisions. The first stage is called *concept screening* and the second stage is called *concept scoring*. Each is supported by a decision matrix which is used by the team to rate, rank, and select the best concept(s). Although the method is structured, we emphasize the role of group insight to improve and combine concepts.

Concept selection is often performed in two stages as a way to manage the complexity of evaluating dozens of product concepts. The application of these two methods is illustrated in Figure 6.4. Screening is a quick, approximate evaluation aimed at producing a few viable alternatives. Scoring is a more careful analysis of these relatively few concepts in order to choose the single concept most likely to lead to product success.

During concept screening, rough initial concepts are evaluated relative to a common reference concept using the *screening matrix.* At this preliminary stage, detailed quantitative comparisons are difficult to obtain and may be misleading, so a coarse comparative rating system is used. After some alternatives are eliminated, the team may choose to move on to concept scoring and conduct more detailed analysis and finer quantitative evaluation of the remaining concepts using the *scoring matrix* as a guide. Throughout the screening and scoring process, several it-

Selection Criteria	Concepts						
	A	**B**	**C**	**D**	**E**	**F**	**G**
	Master Cylinder	**Rubber Brake**	**Ratchet**	**(Reference) Plunge Stop**	**Swash Ring**	**Lever Set**	**Dial Screw**
Ease of handling	0	0	-	0	0	-	-
Ease of use	0	-	-	0	0	+	0
Readability of settings	0	0	+	0	+	0	+
Dose metering accuracy	0	0	0	0	-	0	0
Durability	0	0	0	0	0	+	0
Ease of manufacture	+	-	-	0	0	-	0
Portability	+	+	0	0	+	0	0
Sum +'s	2	1	1	0	2	2	1
Sum 0's	5	4	3	7	4	3	5
Sum –'s	0	2	3	0	1	2	1
Net Score	2	-1	-2	0	1	0	0
Rank	1	6	7	3	2	3	3
Continue?	Yes	No	No	Combine	Yes	Combine	Revise

Figure 6.5

The concept screening matrix. For the syringe example, the team rated the concepts against the reference concept using a simple code (+ for "better than," 0 for "same as," – for "worse than") in order to identify some concepts for further consideration. Note that the three concepts ranked "3" all received the same net score.

erations may be performed, with new alternatives arising from the combination of the features of several concepts. Figures 7-5 and 7-7 illustrate the screening and scoring matrices, using the selection criteria and concepts from the syringe example.

Both stages, concept screening and concept scoring, follow a six-step process which leads the team through the concept selection activity. The steps are:

1. Prepare the selection matrix.
2. Rate the concepts.
3. Rank the concepts.
4. Combine and improve the concepts.
5. Select one or more concepts.
6. Reflect on the results and the process.

Although we present a well-defined process, the team, not the method, creates the concepts and makes the decisions that determine the quality of the product. Ideally, teams are made up of people from different functional groups within the organization. Each member brings unique views that increase the understanding of the problem and thus facilitate the development of a successful, customer-oriented product. The concept selection method exploits the matrices as visual guides for consensus building among team members. The matrices focus attention on the customer needs and other decision criteria and on the product concepts for explicit evaluation, improvement, and selection.

6.5 | CONCEPT SCREENING

Concept screening is based on a method developed by the late Stuart Pugh in the 1980s and is often called *Pugh concept selection* (Pugh, 1990). The purposes of this stage are to narrow the number of concepts quickly and to improve the concepts. Figure 6.5 illustrates the screening matrix used during this stage.

Step 1: Prepare the Selection Matrix

To prepare the matrix, the team selects a physical medium appropriate to the problem at hand. Individuals and small groups with a short list of criteria may use matrices on paper similar to Figure 6.5 or Appendix A for their selection process. For larger groups a chalkboard or flip chart is desirable to facilitate group discussion.

Next, the inputs (concepts and criteria) are entered on the matrix. Although possibly generated by different individuals, concepts should be presented at the same level of detail for meaningful comparison and unbiased selection. The concepts are best portrayed by both a written description and a graphical representation. A simple one-page sketch of each concept greatly facilitates communication of the key features of the concept. The concepts are entered along the top of the matrix, using graphical or textual labels of some kind.

If the team is considering more than about 12 concepts, the *multivote* technique may be used to quickly choose the dozen or so concepts to be evaluated with the screening matrix. Multivoting is a technique in which members of the team simultaneously vote for three to five concepts by applying "dots" to the sheets describing their preferred concepts. The concepts with the most dots are chosen for concept screening. It is also possible to use the screening matrix method with a large number of concepts. This is facilitated by a spreadsheet and it is then useful to transpose the rows and columns. (Arrange the concepts in this case in the left column and the criteria along the top.)

The selection criteria are listed along the left-hand side of the screening matrix, as shown in Figure 6.6. These criteria are chosen based on the customer needs the team has identified, as well as on the needs of the enterprise, such as low manufacturing cost or minimal risk of product liability. The criteria at this stage are usually expressed at a fairly high level of abstraction and typically include from 5 to 10 dimensions. The selection criteria should be chosen to differentiate among the concepts. However, because each criterion is given equal weight in the concept screening method, the team should be careful not to list many relatively unimportant criteria in the screening matrix. Otherwise, the differences among the concepts relative to the more important criteria will not be clearly reflected in the outcome.

After careful consideration, the team chooses a concept to become the benchmark, or *reference concept,* against which all other concepts are rated. The reference is generally either an industry standard or a straightforward concept with which the team members are very familiar. It can be a commercially available product, a best-in-class benchmark product which the team has studied, an earlier generation of the product, any one of the concepts under consideration, or a combination of subsystems assembled to represent the best features of different products.

Step 2: Rate the Concepts

A relative score of "better than" (+), "same as" (0), or "worse than" (-) is placed in each cell of the matrix to represent upon how each concept rates in comparison to the reference concept relative to the particular criterion. It is generally advisable to rate every concept on one criterion before moving to the next criterion. However, with large number of concepts, it is faster to use the opposite approach—to rate each concept completely before moving on to the next concept.

Some people find the coarse nature of the relative ratings difficult to work with. However, at this stage in the design process, each concept is only a general notion of the ultimate product, and more detailed ratings are largely meaningless. In fact, given the imprecision of the concept descriptions at this point, it is very difficult to consistently compare concepts to one another unless one concept (the reference) is consistently used as a basis for comparison.

When available, objective metrics can be used as the basis for rating a concept. For example, a good approximation of assembly cost is the number of parts in a design. Similarly, a good approximation of ease of use is the number of operations required to use the device. These objective metrics help to minimize the judgmental nature of the rating process. Some objective metrics suitable for concept selection may arise from the process of establishing target specifications for the product. (See Chapter 5, Product Specifications, for a discussion of metrics.) Absent objective metrics, ratings are established by team consensus, although secret ballot or other methods may also be useful. At this point the team may also wish to note which selection criteria need further investigation and analysis.

Step 3: Rank the Concepts

After rating all the concepts, the team sums the number of "better than," "same as," and "worse than" scores and enters the sum for each category in the lower rows of the matrix. From our example in Figure 6.6, concept A was rated to have two criteria better than, five the same as, and none worse than the reference concept. Next, a net score can be calculated by subtracting the number of "worse than" ratings from the "better than" ratings.

Once the summation is completed, the team rank-orders the concepts. Obviously, in general those concepts with more pluses and fewer minuses are ranked higher. Often at this point the team can identify one or two criteria which really seem to differentiate the concepts.

Step 4: Combine and Improve the Concepts

Having rated and ranked the concepts, the team should verify that the results make sense and then consider if there are ways to combine and improve certain concepts. Two issues to consider are:

- Is there a generally good concept which is degraded by one bad feature? Can a minor modification improve the overall concept and yet preserve a distinction from the other concepts?
- Are there two concepts which can be combined to preserve the "better than" qualities while annulling the "worse than" qualities?

Combined and improved concepts are then added to the matrix, rated by the team, and ranked along with the original concepts. In our example, the team noticed that concepts D and F could be combined to remove several of the "worse than" ratings to yield a new concept, DF, to be considered in the next round. Concept G was also considered for revision. The team decided that this concept was too bulky, so the excess storage space was removed while retaining the injection technique. These revised concepts are shown in Figure 6.6.

Step 5: Select One or More Concepts

Once the team members are satisfied with their understanding of each concept and its relative quality, they decide which concepts are to be selected for further refine-ment and analysis. During the previous steps, the team will likely develop a clear sense of which are the most promising concepts. The number of concepts selected for further review will be limited by team resources (personnel, money, and time). In our example, the team selected concepts A and E to be considered along with the revised concept G+ and the new concept DF. Having determined the concepts for further analysis, the team must clarify which issues need to be investigated further before a final selection can be made.

The team must also decide whether another round of concept screening will be performed or whether concept scoring will be applied next. If the screening matrix is not seen to provide sufficient resolution for the next step of evaluation and selection, then the concept-scoring stage with its weighted selection criteria and more detailed rating scheme would be used.

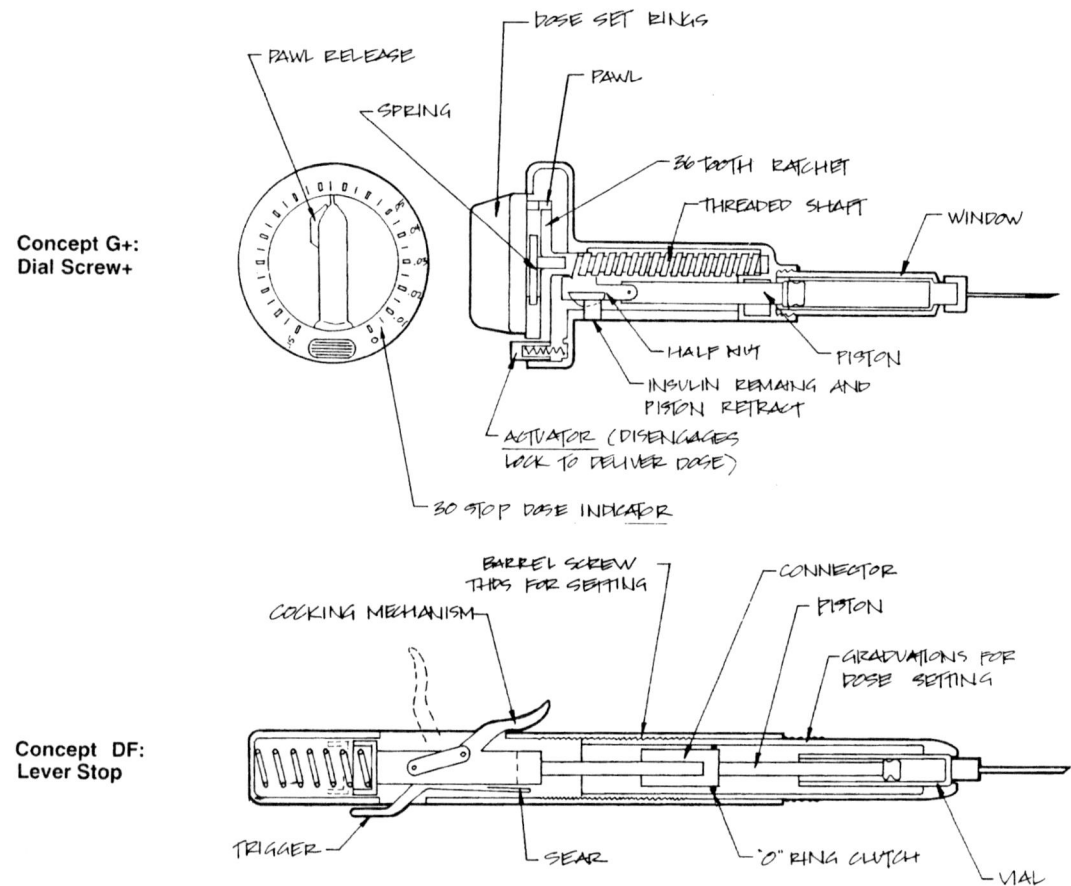

Figure 6.6
New and revised concepts for the syringe. During the selection process, the syringe team revised concept G and generated a new concept, DF, arising from the combination of concepts D and F.

Concepts

selection criteria	weight	a		df (reference) master cylinder		e lever stop		g+ swash ring		dial screw+	
		rating	weighted score	rating	weighted score	rating	weighted score	rating	weighted score	rating	weighted score
ease of handling	5%			**3**	0.15	3	0.15	4	0.2	4	0.2
ease of use	15%			**3**	0.45	4	0.6	4	0.6	3	0.45
readability of settings	10%			2	0.2	**3**	0.3	5	0.5	5	0.5
dose metering accuracy	25%			**3**	0.75	3	0.75	2	0.5	3	0.75
durability	15%			2	0.3	5	0.75	4	0.6	**3**	0.45
ease of manufacture	20%			**3**	0.6	3	0.6	2	0.4	2	0.4
portability	10%			**3**	0.3	3	0.3	3	0.3	3	0.3
	total score			2.75		3.45		3.10		3.05	
	rank			4		1		2		3	
	continue?			no		develop		no		no	

Figure 6.7
The concept scoring matrix. This method uses a weighted sum of the ratings to determine concept ranking. While concept A serves as the overall reference concept, the separate reference points for each criterion are signified by **bold** rating values.

Step 6: Reflect on the Results and the Process

All of the team members should be comfortable with the outcome. If an individual is not in agreement with the decision of the team, then perhaps one or more important criteria are missing from the screening matrix, or perhaps a particular rating is in error, or at least is not clear. An explicit consideration of whether the results make sense to everyone reduces the likelihood of making a mistake and increases the likelihood that the entire team will be solidly committed to the subsequent development activities.

6.6 | CONCEPT SCORING

Concept scoring is used when increased resolution will better differentiate among competing concepts. In this stage, the team weighs the relative importance of the selection criteria and focuses on more refined comparisons with respect to each criterion. The concept scores are determined by the weighted sum of the ratings. Figure 6.7 illustrates the scoring matrix used in this stage. In describing the concept scoring process, we focus on the differences relative to concept screening.

Step 1: Prepare the Selection Matrix

As in the screening stage, the team prepares a matrix and identifies a reference concept. In most cases a computer spreadsheet is the best format to facilitate ranking and sensitivity analysis. The concepts which have been identified for analysis are entered on the top of the matrix. The concepts have typically been refined to some extent since concept screening and may be expressed in more detail. In conjunction with more detailed concepts, the team may wish to add more detail to the selection criteria. The use of hierarchical relations is a useful way to illuminate the criteria. For the syringe example, suppose the team decided that the criterion "ease of use" did not provide sufficient detail to help distinguish among the remaining concepts. "Ease of use" could be broken down, as shown in Figure 6.8, to include "ease of injection," "ease of cleaning," and "ease of loading." The level of criteria detail will depend upon the needs of the team; it may not be necessary to expand the criteria at all. If the team has created a hierarchical list of customer needs, the secondary and tertiary needs are good candidates for more detailed selection criteria. (See Chapter 4, Identifying Customer Needs, for an explanation of primary, secondary, and tertiary needs, and see Appendixes A and B for examples of hierarchical selection criteria.)

After the criteria are entered, the team adds importance weights to the matrix. Several different schemes can be used to weigh the criteria, such as assigning an importance value from 1 to 5, or allocating 100 percentage points among them, as the team has done in Figure 6.7. There are marketing techniques for empirically determining weights from customer data, and a thorough process of identifying

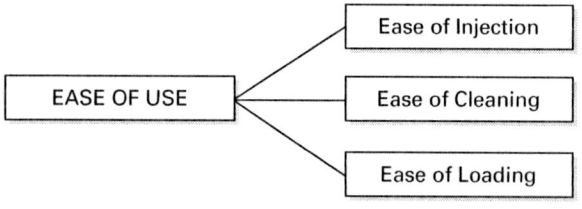

Figure 6.8
Hierarchical decomposition of selection criteria. In conjunction with more detailed concepts, the team may need to break down criteria to the level of detail necessary for meaningful comparison.

customer needs may result in such weights (Urban and Hauser, 1993). However, for the purpose of concept selection the weights are often determined subjectively by team consensus.

Step 2: Rate the Concepts

As in the screening stage, it is generally easiest for the team to focus its discussion by rating all of the concepts with respect to one criterion at a time. Because of the need for additional resolution to distinguish among competing concepts, a finer scale is now used. We recommend a scale from 1 to 5:

Relative Performance	Rating
Much worse than reference	1
Worse than reference	2
Same as reference	3
Better than reference	4
Much better than reference	5

Another scale, such as 1 to 9, may certainly be used, but finer scales generally require more time and effort.

A single reference concept can be used for the comparative ratings, as in the screening stage; however, this is not always appropriate. Unless by pure coincidence the reference concept is of average performance relative to all of the criteria, the use of the same reference concept for the evaluation of each criterion will lead to "scale compression" for some of the criteria. For example, if the reference concept happens to be the easiest concept to manufacture, all of the remaining concepts will receive an evaluation of 1, 2, or 3 ("much worse than," "worse than," or "same as") for the ease-of-manufacture criterion, compressing the rating scale from five levels to three levels.

To avoid scale compression, we recommend using different reference points for the various selection criteria. Reference points may come from several of the concepts

under consideration, from comparative benchmarking analysis, from the target values of the product specifications, or other means. It is important that the reference point for each criterion be well understood to facilitate direct one-to-one comparisons. Using multiple reference points does not prevent the team from designating one concept as the overall reference for the purposes of ensuring that the selected concept is competitive relative to this benchmark. Under such conditions the overall reference concept will simply not receive a neutral score.

Figure 6.7 shows the scoring matrix for the syringe example. The team believed that the master cylinder concept was not suitable as a reference point for two of the criteria, and other concepts were used as reference points in these cases.

Appendix B illustrates a more detailed scoring matrix for which the team rated the concepts on each criterion with no explicit reference points. These ratings were accomplished by discussing the merits of every concept with respect to one criterion at a time, and arranging the scores on a 9-point scale.

Step 3: Rank the Concepts

Once the ratings are entered for each concept, weighted scores are calculated by multiplying the raw scores by the criteria weights. The total score for each concept is the sum of the weighted scores:

$$S_j = \sum_{i-1}^{n} r_{ij} w_i$$

where

r_{ij} = raw rating of concept j for the ith criterion

w_i = weighting for ith criterion

n = number of criteria

S_j = total score for concept j

Finally, each concept is given a rank corresponding to its total score, as shown in Figure 6.7.

Step 4: Combine and Improve the Concepts

As in the screening stage, the team looks for changes or combinations that improve concepts. Although the formal concept generation process is typically completed before concept selection begins, some of the most creative refinements and improvements occur during the concept selection process as the team realizes the inherent strengths and weaknesses of certain features of the product concepts.

Step 5: Select One or More Concepts

The final selection is not simply a question of choosing the concept that achieves the highest ranking after the first pass through the process. Rather, the team should explore this initial evaluation by conducting a sensitivity analysis. Using a computer spreadsheet, the team can vary weights and ratings to determine their effect on the ranking.

By investigating the sensitivity of the ranking to variations in a particular rating, the team members can assess whether uncertainty about a particular rating has a large impact on their choice. In some cases they may select a lower-scoring concept about which there is little uncertainty instead of a higher-scoring concept that may possibly prove unworkable or less desirable as they learn more about it.

Based on the selection matrix, the team may decide to select the top two or more concepts. These concepts may be further developed, prototyped, and tested to elicit customer feedback. See Chapter 8, Concept Testing, for a discussion of methods to assess customer response to product concepts.

The team may also create two or more scoring matrices with different weightings to yield the concept ranking for various market segments with different customer preferences. It may be that one concept is dominant for several segments. The team should also consider carefully the significance of differences in concept scores. Given the resolution of the scoring system, small differences are generally not significant.

For the syringe example, the team agreed that concept DF was the most promising and would be likely to result in a successful product.

Step 6: Reflect on the Results and the Process

As a final step the team reflects on the selected concept(s) and on the concept selection process. In some ways, this is the "point of no return" for the concept development process, so everyone on the team should feel comfortable that all of the relevant issues have been discussed and that the selected concept(s) have the greatest potential to satisfy customers and be economically successful.

After each stage of concept selection, it is a useful reality check for the team to review each of the concepts that are to be eliminated from further consideration. If the team agrees that any of the dropped concepts is better overall than some of those retained, then the source of this inconsistency should be identified. Perhaps an important criterion is missing, not weighted properly, or inconsistently applied.

The organization can also benefit from reflection on the process itself. Two questions are useful in improving the process for subsequent concept selection activities:

- In what way (if at all) did the concept selection method facilitate team decision making?
- How can the method be modified to improve team performance?

These questions focus the team on the strengths and weaknesses of the methodology in relation to the needs and capabilities of the organization.

6.7 CAVEATS

With experience, users of the concept selection methods will discover several subtleties. Here we discuss some of these subtleties and point out a few areas for caution.

- ***Decomposition of concept quality:*** The basic theory underlying the concept selection method is that selection criteria—and, by implication, customer needs—can be evaluated independently and that concept quality is the sum of the qualities of the concept relative to each criterion. The quality of some product concepts may not be easily decomposed into a set of independent criteria, or the performance of the concept relative to the different criteria may be difficult to relate to overall concept quality. For example, the overall appeal or performance of a tennis racquet design may arise in a highly complex way from its weight, ease of swinging, shock transmission, and energy absorption. Simply choosing a concept based on the sum of performance relative to each criterion may fail to capture complex relationships among these criteria. Keeney and Raiffa (1993) discuss the problem of multiattribute decision making, including the issue of nonlinear relationships among selection criteria.
- ***Subjective criteria:*** Some selection criteria, particularly those related to aesthetics, are highly subjective. Choices among alternatives based solely on subjective criteria must be made carefully. In general, the development team's collective judgment is not the best way to evaluate concepts on subjective dimensions. Rather, the team should narrow the alternatives to three or four and then solicit the opinions of representative customers from the target market for the product, perhaps using mock-ups or models to represent the concepts. (See Chapter 8, Concept Testing.)
- ***To facilitate improvement of concepts:*** While discussing each concept to determine its rating, the team

may wish to make note of any outstanding (positive or negative) attributes of the concepts. It is useful to identify any features which could be applied to other concepts, as well as issues which could be addressed to improve the concept. Notes may be placed directly in the cells of the selection matrix. Such notes are particularly useful in step 4, when the team seeks to combine, refine, and improve the concepts before making a selection decision.

• *Where to include cost:* Most of the selection criteria are adaptations of the customer needs. However, "ease of manufacturing" and "manufacturing cost" are not customer needs. The only reason customers care about manufacturing cost is that it establishes the lower bound on sale price. Nevertheless, cost is an extremely important factor in choosing a concept, because it is one of the factors determining the economic success of the product. For this reason, we advocate the inclusion of some measure of cost or ease of manufacturing when evaluating concepts, even though these measures are not true customer needs. Similarly, there may be needs of other stakeholders that were not expressed by actual customers but are important for economic success of the product.

• *Selecting elements of aggregate concepts:* Some product concepts are really aggregations of several simpler concepts. If all of the concepts under consideration include choices from a set of simpler elements, then the simple elements can be evaluated first and in an independent fashion before the more complex concepts are evaluated. This sort of decomposition may follow partly from the structure used in concept generation. For example, if all of the syringes in our example could be used with all of several different needle types, then the selection of a needle concept could be conducted independently of the selection of an overall syringe concept.

• *Applying concept selection throughout the development process:* Although throughout this chapter we have emphasized the application of the method to the selection of a basic product concept, concept selection is used again and again at many levels of detail in the design and development process. For example, in the syringe example, concept selection could be used at the very beginning of the development project to decide between a single-use or multiple-use approach. Once the basic approach had been determined, concept selection could be used to choose the basic product concept, as illustrated in this chapter. Finally, concept selection could be used at the most detailed level of design for resolving decisions such as the choice of colors or materials.

6.8 | SUMMARY

Concept selection is the process of evaluating concepts with respect to customer needs and other criteria, comparing the relative strengths and weaknesses of the concepts, and selecting one or more concepts for further investigation or development.

• All teams use some method, implicit or explicit, for selecting concepts. Decision techniques employed for selecting concepts range from intuitive approaches to structured methods.
• Successful design is facilitated by structured concept selection. We recommend a two-stage process: concept screening and concept scoring.
• Concept screening uses a reference concept to evaluate concept variants against selection criteria. Concept scoring may use different reference points for each criterion.
• Concept screening uses a coarse comparison system to narrow the range of concepts under consideration.
• Concept scoring uses weighted selection criteria and a finer rating scale. Concept scoring may be skipped if concept screening produces a dominant concept.
• Both screening and scoring use a matrix as the basis of a six-step selection process. The six steps are:
  1. Prepare the selection matrix.
  2. Rate the concepts.
  3. Rank the concepts.
  4. Combine and improve the concepts.
  5. Select one or more concepts.
  6. Reflect on the results and the process.
• Concept selection is applied not only during concept development but throughout the subsequent design and development process.
• Concept selection is a group process that facilitates the selection of a winning concept, helps build team consensus, and creates a record of the decision-making process.

6.9 | REFERENCES AND BIBLIOGRAPHY

Many current resources are available on the Internet via

www.ulrich-eppinger.net

The concept selection methodology is a decision-making process. Souder outlines other decision techniques.

Souder, William E., *Management Decision Methods for Managers of Engineering and Research,* Van Nostrand Reinhold Co., New York, NY, 1980.

For a more formal treatment of multiattribute decision making, illustrated with a set of eclectic and interesting case studies, see Keeney and Raiffa.

> Keeney, Ralph L., and Howard Raiffa, *Decisions with Multiple Objectives: Preferences and Value Trade-Offs,* Cambridge University Press, New York, NY, 1993.

Pahl and Beitz's influential engineering design textbook contains an excellent set of systematic methods. The book outlines two concept selection methods similar to concept scoring.

> Pahl, Gerhard, and Wolfgang Beitz, *Engineering Design: A Systematic Approach,* second edition, Ken Wallace (ed.), Springer-Verlag, London, 1996.

Weighting alternatives for selection is not a new idea. The following is one of the earlier references for using selection matrices with weights:

> Alger, J. R., and C. V. Hays, *Creative Synthesis in Design,* Prentice Hall, Englewood Cliffs, NJ, 1964.

The concept-screening method is based upon the concept selection process presented by Stuart Pugh. Pugh was known to criticize more quantitative methods, such as the concept-scoring method presented in this chapter. He cautioned that numbers can be misleading and can re-duce the focus on creativity required to develop better concepts.

> Pugh, Stuart, *Total Design,* Addison-Wesley, Reading, MA, 1990.

Concept scoring is similar to a method often called the Kepner-Tregoe method. It is described, along with other techniques for problem identification and solution, in their text.

> Kepner, Charles H., and Benjamin B. Tregoe, *The Rational Manager,* McGraw-Hill, New York, NY, 1965.

Urban and Hauser describe techniques for determining the relative importance of different product attributes.

> Urban, Glen L., and John R. Hauser, *Design and Marketing of New Products,* second edition, Prentice Hall, Englewood Cliffs, NJ, 1993.

Otto and Wood present a method to include certainty bounds with the ratings given to concepts in concept scoring. These can be combined to derive an estimate of the error in selecting the highest-scoring concept and to compute an confidence interval for the results.

> Otto, Kevin N., and Kristin L. Wood, "Estimating Errors in Concept Selection," ASME Design Engineering Technical Conferences, Vol. DE-83, 1995, pp.397-412.

Exercises

1. How can the concept selection methods be used to benchmark or evaluate existing products? Perform such an evaluation for five automobiles you might consider purchasing.
2. Propose a set of selection criteria for the choice of a battery technology for use in a portable computer.
3. Perform concept screening for the four pencil holder concepts shown below. Assume the pencil holders are for a member of a product development team who is continually moving from site to site.
4. Repeat Exercise 3, but use concept scoring.

Thought Questions

1. How might you use the concept selection method to decide whether to offer a single product to the marketplace or to offer several different product options?
2. How might you use the method to determine which product features should be standard and which should be optional or add-ons?
3. Can you imagine an interactive computer tool that would allow a large group (say, 20 or more people) to participate in the concept selection process? How might such a tool work?
4. What could cause a situation in which a development team uses the concept selection method to agree on a concept that then results in commercial failure?

Appendix A: Concept-Screening Matrix Example

This matrix was created and used by a development team designing a collar to hold weights onto a barbell.

Master Selection Criteria	Velcro Handcuff	Rubber Lock	Alligator Belt	4-Part Belt	Torsional Clip	Screw Latch (REF)	Wing Spring	Type	Hose Nut	Clothespin	Spring-Clamp	Magnetic C-Clamp	Threaded Loaded Bar	Plates	Bar
Functionality															
Lightweight	+	0	+	+	+	0	+	-	-	+	0	0	+	+	0
Fits different bars	+	0	+	+	+	0	0	0	0	+	0	+	0	-	0
Weights secured laterally	0	0	-	-	0	0	0	-	+	-	0	0	-	0	+
Convenience															
Tighten from end/side	0	0	0	0	0	0	-	-	-	0	-	0	+	+	-
Does not roll	0	0	0	0	0	0	0	0	0	0	0	0	0	0	0
Change weights without removing collar	0	0	0	0	0	0	0	0	0	0	0	0	+	+	0
Convenience of placement when changing weights	0	0	+	+	0	0	-	-	-	0	-	0	+	+	-
Ergonomics															
Secure/release (one motion)	+	0	-	-	+	0	-	-	-	0	-	-	+	-	-
Low force to secure/release	0	0	0	0	-	0	-	0	0	0	0	0	+	-	0
RH/LH usage	0	0	+	0	0	0	-	-	0	-	-	0	0	-	-
Not slippery when wet	0	0	+	0	0	0	0	0	0	0	0	0	+	+	0
Use with/one hand	+	0	0	0	+	0	0	0	0	0	0	0	+	+	0
Durability															
Longevity	-	-	-	-	0	0	0	+	0	0	+	+	-	-	+
Other															
Cost of raw materials	0	0	+	+	0	0	0	0	-	+	0	0	-	-	-
Manufacturability	0	-	+	+	0	0	0	+	-	+	+	0	-	-	-
Uses existing weight bars	0	0	0	0	0	0	0	0	0	0	0	0	-	0	-
Sum +'s	4	0	6	4	4	0	2	1	1	2	2	8	6	2	2
Sum 0's	11	14	7	11	11	16	8	8	11	10	12	3	4	7	7
Sum -'s	1	2	3	1	1	0	6	7	4	4	2	5	6	7	7
Net Score	3	-2	3	3	3	0	-4	-6	-3	-2	0	3	0	-5	-5
Rank	1	10	1	1	7	7	13	15	12	10	7	1	7	15	15

Appendix B: Concept-Scoring Matrix Example

A development team generated this matrix while selecting a new concept for a spill-proof beverage holder to be used on boats. Note that in this case the team chose not to define a single concept as the reference for all of the selection criteria.

Selection Criteria	Weight	Concept A Rating	Concept A Weighted Score	Rating	Weighted Score	Rating	Weighted Score	Rating	Weighted Score	Rating	Weighted Score	Rating	Weighted Score	Rating	Weighted Score
Flexible Use	20														
Use in different locations	15	7	105	7	105	8	120	6	90	6	90	5	75	7	105
Holds different beverages	5	5	25	5	25	3	15	4	20	5	25	3	15	3	15
Maintains drink condition	15														
Retains temperature of drink	13	5	65	5	65	5	65	1	13	5	65	5	65	5	65
Prevents water from getting in	2	5	10	7	14	5	10	5	10	5	10	5	10	5	10
Survives boating environment	5														
Doesn't break when dropped	1	6	6	6	6	9	9	7	7	5	5	9	9	6	6
Resists corrosion from sea spray	2	7	14	7	14	8	16	8	16	5	10	9	18	7	14
Floats when it falls in water	2	5	10	6	12	8	16	4	8	5	10	8	16	7	14
Keeps drink container stable	20														
Prevents spilling	7	3	21	4	28	3	21	5	35	5	35	3	21	3	21
Prevents bouncing in waves	6	7	42	8	48	7	42	5	30	5	30	7	42	7	42
Will not slide during pitch/roll	7	5	35	5	35	5	35	5	35	5	35	5	35	5	35
Requires little maintenance	5														
Easily stored when not in use	1	7	7	6	6	8	8	9	9	4	4	8	8	7	7
Easy to maintain a clean appearance	2	6	12	6	12	3	6	4	8	5	10	5	10	6	12
Allows liquid to drain out bottom	2	5	10	5	10	5	10	5	10	5	10	5	10	5	10
Easy to use	15														
Usable with one hand	5	7	35	7	35	7	35	6	30	5	25	7	35	7	35
Easy/comfortable to grip	5	8	40	8	40	6	30	5	25	5	25	6	30	8	40
Easy to exchange beverage containers	2	5	10	5	10	5	10	8	16	5	10	5	10	5	10
Works reliably	3	3	9	3	9	3	9	3	9	4	12	4	12	3	9
Attractive in environment	10														
Doesn't damage boat surface	5	8	40	8	40	8	40	8	40	8	40	6	30	8	40
Attractive to look at	5	7	35	8	40	3	15	4	20	5	25	5	25	8	40
Manufacturing ease	10														
Low-cost materials	4	5	20	4	16	7	28	8	32	4	16	8	32	6	24
Low complexity of parts	3	4	12	3	9	7	21	4	12	3	9	8	24	5	15
Low number of assembly steps	3	5	15	8	24	8	24	3	9	3	9	8	24	6	18
Total Score			578		594		585		484		510		556		587
Rank			4		1		3		7		6		5		2

Part II

The Role of Ethics in Engineering Design

PATH OF
MOTION

Chapter

Engineering as Social Experimentation

7

As it departed on its maiden voyage in April 1912, the *Titanic* was proclaimed the greatest engineering achievement ever. Not merely was it the largest ship the world had seen, having a length of two and a half football fields, it was also the most glamorous of ocean liners, complete with a tropical vinegarden restaurant and the first seagoing masseuse. It was supposed to be the first totally safe ship. Since the worst collision envisaged was at the juncture of 2 of its 16 watertight compartments, and since it could float with any four compartments flooded, the *Titanic* was confidently believed to be virtually unsinkable.

Buoyed by such confidence, the captain allowed the ship to sail full speed at night in an area frequented by icebergs, causing a collision which tore a large gap in the ship's side, directly or indirectly flooding five compartments. Time remained to evacuate the ship, but there were not enough lifeboats to accommodate all the passengers and crew. British regulations then in effect did not foresee vessels of this size. Accordingly, only 825 places were required in lifeboats, sufficient for a mere one-quarter of the *Titanic*'s capacity of 3547 passengers and crew. No extra precautions had seemed necessary for a ship believed to be practically unsinkable. The result: 1522 dead (drowned or frozen) out of the 2227 on board for the Titanic's first trip.[1]

The *Titanic* remains a haunting image of technological complacency. Most products of technology present some potential dangers, and thus engineering is an inherently risky activity. In order to underscore this fact and help in exploring its ethical implications, we suggest that engineering should be viewed as an experimental process. It is not, of course, an experiment conducted solely in a laboratory under controlled conditions. Rather, it is an experiment on a social scale involving human subjects.

Wherever great risk to human life is involved, a ready means of escape (a "safe exit") should be provided. With

this in mind, it should not matter why the *Titanic* sank. There are conjectures that the *Titanic* left England with a coal fire on board, that this made the captain rush the ship to New York, and that water entering the coal bunkers through the gash caused an explosion and greater damage to the compartments. Others maintain that embrittlement of the ship's steel hull in the icy waters caused a much larger crack than a collision would otherwise have produced. Shipbuilders have argued that having the watertight bulkheads reach higher on such a big ship would have kept the ship afloat, but this would have restricted space on the passenger decks for cabins and paying passengers. However, what matters most is that the lack of lifeboats and the difficulty of launching those available from the listing ship prevented safe exit for most persons aboard.

7.1 | ENGINEERING AS EXPERIMENTATION

Experimentation is commonly recognized to play an essential role in the design process. Preliminary tests or simulations are conducted from the time a new design concept is given its first rough design. Materials and processes are tried out, usually employing formal experimental techniques. Such tests serve as the basis for more detailed designs, which in turn are tested. At the production stage, further tests are run, until a finished product evolves. The normal design process is thus iterative, carried out on trial designs with modifications being made on the basis of feedback information acquired from tests. Beyond those specific tests and experiments, however, each engineering project taken as a totality may itself be viewed as an experiment.

Similarities to Standard Experiments

Several features of virtually every kind of engineering practice combine to make it appropriate to view engineering projects as experiments. First, any project is carried

[1] Walter Lord, *A Night to Remember* (New York: Holt, 1976); Wynn C. Wade, *The Titanic: End of a Dream* (New York: Penguin, 1980); Michael Davie, *The Titanic* (London: The Bodley Head, 1986).

out in partial ignorance. There are uncertainties in the abstract model used for the design calculations; there are uncertainties in the precise characteristics of the materials purchased; there are uncertainties in the precision of materials processing and fabrication; there are uncertainties about the nature of the stresses the finished product will encounter. Engineers do not have the luxury of waiting until all the relevant facts are in before commencing work. At some point, theoretical exploration and laboratory testing must be bypassed for the sake of moving ahead on a project. Indeed, one talent crucial to an engineer's success lies precisely in the ability to accomplish tasks safely with only a partial knowledge of scientific laws about nature and society.

To undertake a great work, and especially a work of a novel type, means carrying out an experiment. It means taking up a struggle with the forces of nature without the assurance of emerging as the victor after the first attack.

— Louis Marie Henri Navier (1785–1836), bridge builder, founder of structural analysis

---

Second, the final outcomes of engineering projects, like those of experiments, are generally uncertain. Often in engineering, it is not even known what the possible outcomes may be, and great risks may attend even seemingly benign projects. A reservoir may do damage to a region's social fabric or to its ecosystem. It may not even serve its intended purpose if the dam leaks or breaks. An aqueduct in a region where it is the only source of water, may bring about a population explosion, creating dependency and vulnerability without adequate safeguards. A jumbo airplane may bankrupt the small airline that bought it as a status symbol. A special-purpose fingerprint reader may find its main application in the identification and surveillance of dissidents by totalitarian regimes. A nuclear reactor, the scaled-up version of a successful smaller model, may exhibit unexpected problems that endanger the surrounding population, leading to its untimely shutdown at great cost to owner and consumers alike. In the past, a hair dryer may have exposed the unwary user to lung damage from the asbestos insulation in its barrel.

Third, effective engineering relies on knowledge gained about products both before and after they leave the factory—knowledge needed for improving current products and creating better ones. That is, ongoing success in engineering depends upon gaining new knowledge, just as does ongoing success in experimentation. Monitoring is thus as essential to engineering as it is to experimentation

in general. To monitor is to make periodic observations and tests in order to check for both successful performance and unintended side effects. But since the ultimate test of a product's efficiency, safety, cost-effectiveness, environmental impact, and aesthetic value lies in how well that product functions within society, monitoring cannot be restricted to the in-house development or testing phases of an engineering venture. Monitoring should also extend to the stage of client use, because just as in experimentation, both the intermediate and final results of an engineering project deserve analysis if the correct lessons are to be learned from it.

Learning from the Past

t might be expected that engineers would learn not only from their own earlier design and operating results, but also from those of other engineers. Unfortunately, that is frequently not the case. Lack of established channels of communication, misplaced pride in not asking for information, embarrassment at failure or fear of litigation, and plain neglect often impede the flow of such information and lead to many repetitions of past mistakes. Here are a few examples:

1. The *Titanic* lacked a sufficient number of lifeboats decades after most of the passengers and crew on the steamship *Arctic* had perished because of the same problem.[2]

2. "Complete lack of protection against impact by shipping caused Sweden's worst ever bridge collapse on Friday as a result of which eight people were killed." Thus reported the *New Civil Engineer* on January 24, 1980. On May 15 of the same year, it also reported the following: "Last Friday's disaster at Tampa Bay, Florida, was the largest and most tragic of a growing number of incidents of errant ships colliding with bridges over navigable waterways." While collisions of ships with bridges do occur—other well-known cases being those of the Maracaibo Bridge (Venezuela, 1964) and the Tasman Bridge (Australia, 1975)—Tampa's Sunshine Skyline Bridge was not designed with horizontal impact forces in mind because the code did not require them. Some engineers have proposed the use of floating concrete bumpers that can deflect ships.

3. In June 1966, a section of the Milford Haven bridge in Wales collapsed during construction. In

---

[2] Wade, *The Titanic*, p. 417.

October of the same year, a bridge of similar design was being erected by the same bridge-builder (Freeman Fox and Partners) in Melbourne, Australia, when it too partially collapsed, killing 33 people and injuring 19. This happened shortly after chief construction engineer Jack Hindshaw (also a casualty) had assured worried workers that the construction site was safe.[3]

4. Valves are notorious for being among the least reliable components of hydraulic systems. It was a pressure relief valve, and lack of definitive information regarding its open or shut state, that contributed to the nuclear reactor accident at Three Mile Island on March 28, 1979. Similar malfunctions had occurred with identical valves on nuclear reactors at other locations. The required reports had been filed with Babcock and Wilcox, the reactor's manufacturer, but no attention had been given to them.[4]

These examples, and others to be given in later chapters, illustrate why it is not sufficient for engineers to rely on handbooks and computer programs without knowing the limits of the tables and algorithms underlying their favorite tools. They need to visit shop floors and construction sites to learn from workers and foremen how earlier projects have fared during erection or assembly and tests, and how satisfied the customers were. The art of back-of-the-envelope calculations to obtain ball-park values with which to independently check more lengthy and complicated procedures must not be lost. Engineering, just like experimentation, demands practitioners who remain alert and well-informed at every stage of a project's history, and who exchange ideas freely with colleagues in related departments.

Contrasts with Standard Experiments

To be sure, engineering differs in some respects from standard experimentation. Some of those very differences help to highlight the engineer's special responsibilities. And exploring the differences can also aid our thinking about the moral responsibilities of all those engaged in engineering.

Experimental Control   One great difference arises with experimental control. In a standard experiment, this involves the selection, at random, of members for two different groups. The members of one group receive the special, experimental treatment. Members of the other group, called the control group, do not receive that special treatment, although they are subjected to the same environment as the first group in every other respect.

In engineering, this is not the usual practice—unless the project is confined to laboratory experimentation—because the experimental subjects are human beings out of the experimenter's control. Indeed, clients and consumers exercise most of the control because it is they who choose the product or item they wish to use. This makes it impossible to obtain a random selection of participants from various groups. Nor can parallel control groups be established based on random sampling. Thus, it is not possible to study the effects that changes in variables have on two or more comparison groups, and one must simply work with the available historical and retrospective data about various groups that use the product.

This suggests that the view of engineering as social experimentation involves a somewhat extended usage of the concept of experimentation. Nevertheless, "engineering as social experimentation" should not be dismissed as a merely metaphorical notion. There are other fields where it is not uncommon to speak of experiments whose original purpose was not experimental in nature and that involve no control groups.

For example, social scientists monitor and collect data on differences and similarities between existing educational systems that were not initially set up as systematic experiments. In doing so, they regard the current diversity of systems as constituting what has been called a "natural experiment" (as opposed to a deliberately initiated one).[5] Similarly, we think that engineering can be appropriately viewed as just such a "natural experiment" using human subjects, despite the fact that most engineers do not currently consider it in that light.

Informed Consent   Viewing engineering as an experiment on a societal scale places the focus where it should be: on the human beings affected by technology, for the experiment is performed on persons, not on inanimate objects. In this respect, albeit on a much larger scale, engineering closely parallels medical testing of new drugs and techniques on human subjects.

Society has recently come to recognize the primacy of the subject's safety and freedom of choice as to whether to

---

[3] "Yarrow Bridge," editorial, *The Engineer* 210 (October 1970), p. 415.
[4] Robert Sugarman, "Nuclear Power and the Public Risk," *IEEE Spectrum* 16 (November 1979), p. 72.

[5] Alice M. Rivlin, *Systematic Thinking for Social Action* (Washington, DC: The Brookings Institution, 1971), p. 70.

participate in medical experiments. Ever since the revelations of prison and concentration camp horrors in the name of medicine, an increasing number of moral and legal safeguards have arisen to ensure that subjects in experiments participate on the basis of informed consent.

While current medical practice has increasingly tended to accept as fundamental the subject's moral and legal rights to give informed consent before participating in an experiment, contemporary engineering practice is only beginning to recognize those rights. We believe that the problem of informed consent, which is so vital to the concept of a properly conducted experiment involving human subjects, should be the keystone in the interaction between engineers and the public. We are talking about the lay public. When a manufacturer sells a new device to a knowledgeable firm that has its own engineering staff, there is usually an agreement regarding the shared risks and benefits of trying out the technological innovation.

Informed consent is understood as including two main elements: knowledge and voluntariness. First, subjects should be given not only the information they request, but all the information needed to make a reasonable decision. Second, subjects must enter into the experiment without being subjected to force, fraud, or deception. Respect for the fundamental rights of dissenting minorities and compensation for harmful effects are taken for granted here.

The mere purchase of a product does not constitute informed consent, any more than does the act of showing up on the occasion of a medical examination. The public and clients must be given information about the practical risks and benefits of the process or product in terms they can understand. Supplying complete information is neither necessary nor in most cases possible. In both medicine and engineering, there may be an enormous gap between the experimenter's and the subject's understanding of the complexities of an experiment. But while this gap most likely cannot be closed, it should be possible to convey all pertinent information needed for making a reasonable decision on whether to participate.

We do not propose a proliferation of lengthy environmental impact reports. We favor the kind of sound advice a responsible physician gives a patient when prescribing a course of drug treatment that has possible side effects. The physician must search beyond the typical sales brochures from drug manufacturers for adequate information; hospital management must allow the physician the freedom to undertake different treatments for different patients, as each case may constitute a different "experiment" involving different circumstances; finally, the patient must be readied to receive the information.

Likewise, an engineer cannot succeed in providing essential information about a project or product unless there is cooperation by management and also receptivity on the part of those who should have the information. Management is often understandably reluctant to provide more information than current laws require, fearing disclosure to potential competitors and exposure to potential lawsuits. Moreover, it is possible that, paralleling the experience in medicine, clients or the public may not be interested in all of the relevant information about an engineering project, at least not until a crisis looms. It is important nevertheless that all avenues for disseminating such information be kept open and ready.

We note that the matter of informed consent is surfacing indirectly in the continuing debate over acceptable forms of energy. Representatives of the nuclear industry can be heard expressing their impatience with critics who worry about reactor malfunction while engaging in statistically more hazardous activities such as driving automobiles and smoking cigarettes. But what is being overlooked by those industry representatives is the common enough human readiness to accept *voluntarily undertaken risks* (as in daring sports), even while objecting to *involuntary risks* resulting from activities in which the individual is neither a direct participant nor a decision maker. In other words, we all prefer to be the subjects of our own experiments rather than those of somebody else. When it comes to approving a nearby oil-drilling platform or a nuclear plant, affected parties expect their consent to be sought no less than it is when a doctor contemplates surgery.

Prior consultation of the kind suggested can be effective. When Northern States Power Company (Minnesota) was planning a new power plant, it got in touch with local citizens and environmental groups before it committed large sums of money to preliminary design studies. The company was able to present convincing evidence regarding the need for a new plant and then suggested several sites. Citizen groups responded with a site proposal of their own. The latter was found acceptable by the company. Thus, informed consent was sought from and voluntarily given by those the project affected, and the acrimonious and protracted battles so common in other cases where a company has already invested heavily in decisions based on engineering studies alone was avoided.[6]

[6] Peter Borrelli, Mahlon Easterling, Burton H. Klein et al., *People, Power and Pollution* (Pasadena, CA: Environmental Quality Lab, California Institute of Technology, 1971), pp. 36–39.

Note that the utility company interacted with groups that could serve as proxy for various segments of the rate-paying public. Obviously it would have been difficult to involve the rate-payers individually.

We endorse a broad notion of informed consent, or what some would call *valid consent,* defined by the following conditions:[7]

1. The consent was given voluntarily.
2. The consent was based on the information that a rational person would want, together with any other information requested, presented in understandable form.
3. The consenter was competent (not too young or mentally ill, for instance) to process the information and make rational decisions.

We suggest two requirements for situations in which the subject cannot be readily identified as an individual:

4. Information that a rational person would need, stated in understandable form, has been widely disseminated.
5. The subject's consent was offered in proxy by a group that collectively represents many subjects of like interests, concerns, and exposure to risk.

Knowledge Gained

Scientific experiments are conducted to gain new knowledge, while "engineering projects are experiments that are not necessarily designed to produce very much knowledge," according to a valuable interpretation of our paradigm by Taft Broome.[8] When we carry out an engineering activity as if it were an experiment, we are primarily preparing ourselves for unexpected outcomes. The best outcome in this sense is one that tells us nothing new but merely affirms that we are right about something. Unexpected outcomes send us on a search for new knowledge—possibly involving an experiment of the first (scientific) type. For the purposes of our model, the distinction is not vital because we are concerned about the manner in which the experiment is conducted, such as that valid consent of human subjects is sought, safety measures are taken, and means exist for terminating the

experiment at any time with all participants having access to a safe exit.

Discussion Topics

1. On June 5, 1976, Idaho's Teton Dam collapsed, killing 11 people and causing $400 million in damage. The Bureau of Reclamation, which built the ill-fated Teton Dam, allowed it to be filled rapidly, thus failing to provide sufficient time to monitor for the presence of leaks in a project constructed with less-than-ideal soil.[9]

   Drawing upon the concept of engineering as social experimentation, discuss the following facts uncovered by the General Accounting Office and reported in the press.

   (a) Because of the designers' confidence in the basic design of Teton Dam, it was believed that no significant water seepage would occur. Thus, sufficient instrumentation to detect water erosion was not installed.

   (b) Significant information suggesting the possibility of water seepage was acquired at the dam site six weeks before the collapse. It was sent through routine channels from the project supervisors to the designers, and arrived at the designers the day after the collapse.

   (c) During the important stage of filling the reservoir, there was no around-the-clock observation of the dam. As a result, the leak was detected only five hours before the collapse. Even then, the main outlet could not be opened to prevent the collapse because a contractor was behind schedule in completing the outlet structure.

   (d) Ten years earlier, the Bureau's Fontenelle Dam had experienced massive leaks that caused a partial collapse, an experience on which the Bureau could have drawn.

2. Read about the catastrophic failure of a dam (other than the Teton Dam), then prepare a list of procedural questions that designers, builders, and operators of dams as well as local public safety officers and nearby neighbors of dams should be able to answer. (You may wish to consult Why Buildings Fail by M. Levy and M. Salvadori [New York: W. W. Nor-

[7] Charles M. Culver and Bernard Gert, "Valid Consent," in *Conceptual and Ethical Problems in Medicine and Psychiatry,* ed. Charles M. Culver and Bernard Gert (New York: Oxford University Press, 1982).

[8] Taft H. Broome Jr., "Engineering Responsibility for Hazardous Technologies," *Journal of Professional Issues in Engineering* 113 (April 1987), pp. 139–49.

[9] Gaylord Shaw, "Bureau of Reclamation Harshly Criticized in New Report on Teton Dam Collapse," *Los Angeles Times,* June 4, 1977, Part I, p. 3; Philip M. Boffey, "Teton Dam Verdict: Foul-up by the Engineers," *Science* 195 (January 1977), pp. 270–72.

ton & Co., 1987] for cases such as the 1889 Johnstown flood in Pennsylvania and the 1959 collapse of Malpasset Dam in France. Another good source is *Dams and Public Safety* by R. B. Jansen [Denver: USID, Government Printing Office, 1980].)

3. The University of California uses tax dollars to develop farm machinery such as tomato, lettuce, and melon harvesters and fruit tree shakers. Such machinery reduces the need for farm labor and raises farm productivity. It definitely benefits the growers. It is also said to benefit all of society. Farm workers, however, claim that replacing an adequate and willing workforce with machines will generate social costs not offset by higher productivity. Among the costs they cite are the need to retrain farm workers for other jobs and the loss of small farms. Discuss if and how continuing farm mechanization may be viewed as an experiment.

4. Models often influence thinking by effectively organizing and guiding reflection and crystallizing attitudes. Yet they usually have limitations and can themselves be misleading to some degree. Write a short essay in which you critically assess the strengths and weaknesses you see in the social experimentation model.

   One possible criticism you might consider is whether the model focuses too much on the creation of new products, whereas a great deal of engineering involves the routine application of results from past work and projects. Another point to consider is how informed consent is to be measured in situations where groups are involved, as in the construction of a garbage incinerator near a community of people having mixed views about the advisability of constructing the incinerator.

5. Debates over responsibility for safety in regard to technological products often turn on who should be considered mainly responsible: the consumer ("buyer beware") or the manufacturer ("seller beware"). How might an emphasis on the idea of informed consent influence thinking about this question?

6. During 1994, the U.S. government released information that in the decades following World War II, some of its agencies conducted tests on the effects of radiation on unsuspecting civilians. Discuss such practices in the light of secrecy imposed by national security considerations and of the right of subjects of experimentation to be informed of the nature of the tests and their possible effects.

7.2 | ENGINEERS AS RESPONSIBLE EXPERIMENTERS

What are the responsibilities of engineers to society? Viewing engineering as social experimentation does not by itself answer this question. While engineers are the main technical enablers or facilitators, they are far from being the sole experimenters. Their responsibility is shared with management, the public, and others. Yet their expertise places them in a unique position to monitor projects, to identify risks, and to provide clients and the public with the information needed to make reasonable decisions. We want to know what is involved in displaying the virtue of being a responsible person while acting as an engineer. From the perspective of engineering as social experimentation, what are the general features of morally responsible engineers?

At least four elements are pertinent—a conscientious commitment to live by moral values, a comprehensive perspective, autonomy, and accountability[10]—or, stated in greater detail as applied to engineering projects conceived as social experiments:

1. A primary obligation to protect the safety of human subjects and respect their right of consent.

2. A constant awareness of the experimental nature of any project, imaginative forecasting of its possible side effects, and a reasonable effort to monitor them.

3. Autonomous, personal involvement in all steps of a project.

4. Accepting accountability for the results of a project.

It is implied in the foregoing that engineers should also display technical competence and other attributes of professionalism. Inclusion of these four requirements as part of engineering practice would then earmark a definite "style" of engineering. In elaborating upon this style, we will note some of the contemporary threats to it.

Conscientiousness

People act responsibly to the extent that they conscientiously commit themselves to live according to moral values. But moving beyond this truism leads immediately to controversy over the precise nature of those values. Moral values transcend a consuming preoccupation with a nar-

---

[10] Graham Haydon, "On Being Responsible," *The Philosophical Quarterly* 28 (1978), pp. 46–57.

rowly conceived self-interest. Accordingly, individuals who think solely of their own good to the exclusion of the good of others are not moral agents. By conscientious moral commitment we mean a sensitivity to the full range of moral values and responsibilities relevant to a given situation, and the willingness to develop the skill and expend the effort needed to reach the best balance possible among those considerations. It will be noted that conscientiousness implies consciousness (in the sense of awareness), because intent is not sufficient. Open eyes, open ears, and an open mind are required to recognize a given situation, its implications, and who is involved or affected.

The contemporary working conditions of engineers tend to narrow moral vision solely to the obligations that accompany employee status. As stated earlier, some 90 percent of engineers are salaried employees, most of whom work within large bureaucracies under great pressure to function smoothly within the organization. There are obvious benefits in terms of prudent self-interest and concern for one's family that make it easy to emphasize as primary the obligations to one's employer. Gradually, the minimal negative duties, such as not falsifying data, not violating patent rights, and not breaching confidentiality, may come to be viewed as the full extent of moral aspiration.

Conceiving engineering as social experimentation restores the vision of engineers as guardians of the public interest, whose professional duty it is to guard the welfare and safety of those affected by engineering projects. And this helps to ensure that such safety and welfare will not be disregarded in the quest for new knowledge, the rush for profits, a narrow adherence to rules, or a concern over benefits for the many that ignores harm to the few.

The role of social guardian should not suggest that engineers force, paternalistically, their own views of the social good upon society. For, as with medical experimentation on humans, the social experimentation involved in engineering should be restricted by the participant's consent—voluntary and informed consent.

Relevant Information

Conscientiousness is blind without relevant factual information. Hence, showing moral concern involves a commitment to obtain and properly assess all available information pertinent to meeting one's moral obligations. This means, as a first step, fully grasping the context of one's work, which makes it count as an activity having a moral import.

For example, there is nothing wrong in itself with trying to design a good heat exchanger. But if I ignore the fact that the heat exchanger will be used in the manufac-

ture of a potent, illegal hallucinogen, I am showing a lack of moral concern. It is this requirement that one be aware of the wider implications of one's work that makes participation in, say, a design project for a super-weapon morally problematic—and that makes it sometimes convenient for engineers self-deceivingly to ignore the wider context of their activities, a context that may rest uneasily with an active conscience.

Another way of blurring the context of one's work results from the ever-increasing specialization and division of labor that makes it easy to think of someone else in the organization as responsible for what otherwise might be a bothersome personal problem. For example, a company may produce items with obsolescence built into them, or the items might promote unnecessary energy usage. It is easy to place the burden on the sales department: "Let them inform the customers—if the customers ask." It may be natural to thus rationalize one's neglect of safety or cost considerations, but it shows no moral concern. More convenient is a shifting of the burden to the government and voters: "We will attend to this when the government sets standards so our competitors must follow suit," or "Let the voters decide on the use of super-weapons; we just built them."

These ways of losing perspective on the nature of one's work also hinder one in acquiring a full perspective along a second dimension of factual information: the consequences of what one does. And so, while regarding engineering as social experimentation points out the importance of context, it also urges the engineer to view his or her specialized activities in a project as part of a larger whole having a social impact—an impact that may involve a variety of unintended effects. Accordingly, it emphasizes the need for wide training in disciplines related to engineering and its results, as well as the need for a constant effort to imaginatively foresee dangers.

It might be said that the goal is to practice what Chauncey Starr once called "defensive engineering." Or perhaps more fundamental is the need for "preventive technology," which parallels what Ruth Davis says about preventive medicine:

> The solution to the problem is not in successive cures to successive science-caused problems; it is in their prevention. Unfortunately, cures for scientific ills are generally more interesting to scientists than is the prevention of those ills. We have the unhappy history of the medical community to show us the difficulties associated with trying to establish preventive medicine as a specialty.[11]

[11] Ruth M. Davis, "Preventative Technology: A Cure for Specific Ills," *Science* 188 (April 1975), p. 213.

No amount of disciplined and imaginative foresight, however, can serve to anticipate all dangers. Because engineering projects are inherently experimental in nature, it is crucial for them to be monitored on an ongoing basis from the time they are put into effect. While individual practitioners cannot privately conduct full-blown environmental and social impact studies, they can choose to make the extra effort needed to keep in touch with the course of a project after it has officially left their hands. This is a mark of personal identification with one's work, a notion that leads to the next aspect of moral responsibility.

Moral Autonomy

People are morally autonomous when their moral conduct and principles of action are their own, in a special sense derived from Kant: Moral beliefs and attitudes should be held on the basis of critical reflection rather than passive adoption of the particular conventions of one's society, church, or profession. This is often what is meant by "authenticity" in one's commitment to moral values.

Those beliefs and attitudes, moreover, must be integrated into the core of an individual's personality in a manner that leads to committed action. They cannot be agreed to abstractly and formally, and adhered to merely verbally. Thus, just as one's principles are not passively absorbed from others when one is morally autonomous, so too one's actions are not treated as something alien and apart from oneself.

It is a comfortable illusion to think that in working for an employer, and thereby performing acts directly serving a company's interests, one is no longer morally and personally identified with one's actions. Selling one's labor and skills may make it seem that one has thereby disowned and forfeited power over one's actions.[12]

Viewing engineering as social experimentation can help one overcome this tendency and can help restore a sense of autonomous participation in one's work. As an experimenter, an engineer is exercising the sophisticated training that forms the core of his or her identity as a professional. Moreover, viewing an engineering project as an experiment that can result in unknown consequences should help inspire a critical and questioning attitude about the adequacy of current economic and safety stan-

dards. This also can lead to a greater sense of personal involvement with one's work.

The attitude of management plays a decisive role in how much moral autonomy engineers feel they have. It would be in the long-term interest of a high-technology firm to grant its engineers a great deal of latitude in exercising their professional judgment on moral issues relevant to their jobs (and, indeed, on technical issues as well). But the yardsticks by which a manager's performance is judged on a quarterly or yearly basis often discourage this. This is particularly true in our age of conglomerates, when near-term profitability is more important than consistent quality and long-term retention of satisfied customers.

In government-sponsored projects, it is frequently a deadline that becomes the ruling factor, along with fears of interagency or foreign competition. Tight schedules contributed to the loss of the space shuttle *Challenger,* as we shall see later.

Accordingly, engineers are compelled to look to their professional societies and other outside organizations for moral support. Yet it is no exaggeration to claim that the blue-collar worker with union backing has greater leverage at present in exercising moral autonomy than do many employed professionals. Professional societies, originally organized as learned societies dedicated to the exchange of technical information, lack comparable power to protect their members, although most engineers have no other group on which to rely for such protection. Only now is the need for moral and legal support of members in the exercise of their professional obligations being recognized by those societies.[13]

Accountability

Finally, responsible people accept moral responsibility for their actions. Too often, "accountable" is understood in the overly narrow sense of being culpable and blameworthy for misdeeds. But the term more properly refers to the general disposition of being willing to submit one's actions to moral scrutiny and be open and responsive to the assessments of others. It involves a willingness to present morally cogent reasons for one's conduct when called upon to do so in appropriate circumstances.

Submission to an employer's authority, or any authority for that matter, creates in many people a narrowed sense of accountability for the consequences of their actions. This was documented by some famous experiments

---

[12] John Lachs, "'I Only Work Here': Mediation and Irresponsibility," in *Ethics, Free Enterprise, and Public Policy,* ed. Richard T. DeGeorge and Joseph A. Pichler (New York: Oxford, 1978), pp. 201–13. Also see Elizabeth Wolgast, *Ethics of an Artificial Person: Lost Responsibility in Professions and Organizations* (Stanford: Stanford University Press, 1992).

[13] Stephen H. Unger, *Controlling Technology: Ethics and the Responsible Engineer,* 2nd ed. (New York: John Wiley & Sons, 1994).

conducted by Stanley Milgram during the 1960s.[14] Subjects would come to a laboratory believing they were to participate in a memory and learning test. In one variation, two other people were involved, the "experimenter" and the "learner." The experimenter was regarded by the subject as an authority figure, representing the scientific community. He or she would give the subject orders to administer electric shocks to the "learner" whenever the latter failed in the memory test. The subject was told the shocks were to be increased in magnitude with each memory failure. All this, however, was a deception—a setup. There were no real shocks and the apparent "learner" and the "experimenter" were merely acting parts in a ruse designed to see how far the unknowing experimental subject was willing to go in following orders from an authority figure.

The results were astounding. When the subjects were placed in an adjoining room separated from the "learner" by a shaded glass window, more than half were willing to follow orders to the full extent: giving the maximum electric jolt of 450 volts. This was in spite of seeing the "learner," who was strapped in a chair, writhing in (apparent) agony. The same results occurred when the subjects were allowed to hear the (apparently) pained screams and protests of the "learner," screams and protests that became intense from 130 volts on. There was a striking difference, however, when subjects were placed in the same room within touching distance of the "learner." Then the number of subjects willing to continue to the maximum shock dropped by one-half.

Milgram explained these results by citing a strong psychological tendency in people to be willing to abandon personal accountability when placed under authority. He saw his subjects ascribing all initiative, and thereby all accountability, to what they viewed as legitimate authority. And he noted that the closer the physical proximity, the more difficult it becomes to divest oneself of personal accountability.

The divorce between causal influence and moral accountability is common in business and the professions, and engineering is no exception. Such a psychological schism is encouraged by several prominent features of contemporary engineering practice.

First, large-scale engineering projects involve fragmentation of work. Each person makes only a small contribution to something much larger. Moreover, the final product is often physically removed from one's immediate workplace, creating the kind of "distancing" that Milgram identified as encouraging a lessened sense of personal accountability.

Second, corresponding to the fragmentation of work is a vast diffusion of accountability within large institutions. The often massive bureaucracies within which so many engineers work are bound to diffuse and delimit areas of personal accountability within hierarchies of authority.

Third, there is frequently pressure to move on to a new project before the current one has been operating long enough to be observed carefully. This promotes a sense of being accountable only for meeting schedules.

Fourth, the contagion of malpractice suits currently afflicting the medical profession is carrying over into engineering. With this comes a crippling preoccupation with legalities, a preoccupation that makes one wary of becoming morally involved in matters beyond one's strictly defined institutional role.

We do not mean to underestimate the very real difficulties these conditions pose for engineers who seek to act as morally accountable people on their jobs. Much less do we wish to say engineers are blameworthy for all the harmful side effects of the projects on which they work, even though they partially cause those effects simply by working on the projects. That would be to confuse accountability with *blameworthiness,* and also to confuse *causal* responsibility with *moral* responsibility. But we do claim that engineers who endorse the perspective of engineering as a social experiment will find it more difficult to divorce themselves psychologically from personal responsibility for their work. Such an attitude will deepen their awareness of how engineers daily cooperate in a risky enterprise in which they exercise their personal expertise toward goals they are especially qualified to attain, and for which they are also accountable.

Babylon's Building Code, 1758 B.C.

Hammurabi, as king of Babylon, was concerned with strict order in his realm, and he decided that the builders of his time should also be governed by his laws. Thus he decreed:

> If a builder has built a house for a man and has not made his work sound, and the house which he has built has fallen down and so caused the death of the householder, that builder shall be put to death. If it causes the death of the householder's son, they shall put that builder's son to death. If it causes the death of the householder's slave, he shall give slave for slave to the householder. If it destroys property he shall replace anything it has destroyed; and because he has not made sound the house which he has built and it has fallen down, he shall rebuild the house

[14] Stanley Milgram, *Obedience to Authority* (New York: Harper & Row, 1974).

which has fallen down from his own property. If a builder has built a house for a man and does not make his work perfect and the wall bulges, that builder shall put that wall into sound condition at his own cost.[15]

A Balanced Outlook on Law

What should be the role of law in engineering, as viewed within our model of social experimentation? The legal regulations that apply to engineering and other professions are becoming more numerous and more specific all the time. We hear many complaints about this trend, and a major effort to deregulate various spheres of our lives is currently under way. Nevertheless, we continue to hear cries of "there ought to be a law" whenever a crisis occurs or a special interest is threatened.

This should not be surprising to us in the United States. We pride ourselves on being a nation that lives under the rule of law. We even delegate many of our decisions on ethical issues to an interpretation of laws. And yet this emphasis on law can cause problems in regard to ethical conduct beyond more practical issues usually cited by those who favor deregulation.

For example, one of the greatest moral problems in engineering, and one fostered by the very existence of minutely detailed rules, is that of *minimal compliance*. This can find its expression when companies or individuals search for loopholes in the law that will allow them to barely keep to its letter even while violating its spirit. Or hard-pressed engineers find it convenient to refer to standards with ready-made specifications as a substitute for original thought, perpetuating the "handbook mentality" and the repetition of mistakes. Minimal compliance led to the tragedy of the *Titanic*: Why should that ship have been equipped with enough lifeboats to accommodate all its passengers and crew when British regulations at the time required only a lower minimum, albeit with smaller ships in mind?

On the other hand, remedying the situation by continually updating laws or regulations with further specifications may also be counterproductive. Not only will the law inevitably lag behind changes in technology and produce a judicial vacuum, there is also the danger of overburdening the rules and the regulators. As Robert Kates puts it:

If cooperation is not forthcoming—if the manufacturer, for example, falsifies or fails to conduct safety tests—there is

something akin to the law of infinite regress in which the regulator must intrude more and more expensively into the data collection and evaluation process. In the end, the magnitude of the task overwhelms the regulators.[16]

Lawmakers cannot be expected always to keep up with technological development. Nor would we necessarily want to see laws changed upon each innovation. Instead we empower rule-making and inspection agencies—the Food and Drug Administration (FDA), the Federal Aviation Agency (FAA), and the Environmental Protection Agency (EPA) are examples of these in the United States—to fill the void. Though they are independent and belong to neither the judicial nor the executive branches of government, their rules have, for all practical purposes, the effect of law.

Industry tends to complain that excessive restrictions are imposed on it by regulatory agencies. But one needs to reflect on why regulations may have been necessary in the first place. Take, for example, the U.S. Consumer Product Safety Commission's rule for baby cribs, which specifies that "the distance between components (such as slats, spindles, crib rods, and corner posts) shall not be greater than $2^3/_8$ inches at any point." This rule came about because some manufacturers of baby furniture had neglected to consider the danger of babies strangling in cribs or had neglected to measure the size of babies' heads.[17]

Again, why must regulations be so specific when broad statements would appear to make more sense? When the EPA adopted rules for asbestos emissions in 1971, it was recognized that strict numerical standards would be impossible to promulgate. Asbestos dispersal and intake, for example, are difficult to measure in the field. So, being reasonable, the EPA specified a set of work practices to keep emissions to a minimum—that asbestos should be wetted down before handling, for example, and disposed of carefully. The building industry called for more specifics. Modifications in the Clean Air Act eventually permitted EPA to issue enforceable rules on work practices, and later the Occupational Safety and Health Administration also became involved.

Society's attempts at regulation have indeed often failed, but it would be wrong to write off rule-making and rule-following as futile. Good laws, effectively enforced, clearly produce benefits. They authoritatively establish

---

[15] Hammurabi, *The Code of Hammurabi*, trans. R. F. Harper (Chicago: University of Chicago Press, 1904).

[16] Robert W. Kates, ed., *Managing Technological Hazards: Research Needs and Opportunities* (Boulder, CO: Institute of Behavioral Science, University of Colorado, 1977), p. 32.

[17] William W. Lowrance, *Of Acceptable Risk* (Los Altos, CA: William Kaufmann, 1976), p. 134.

reasonable minimal standards of professional conduct and provide at least a self-interested motive for most people and corporations to comply. Moreover, they serve as a powerful support and defense for those who wish to act ethically in situations where ethical conduct might be less than welcome. By being able to point to a pertinent law, one can feel more free to act as a responsible engineer.

Engineering as social experimentation can provide engineers with a proper perspective on laws and regulations in that rules that govern engineering practice should not be devised or construed as rules of a game but as rules of responsible experimentation. Such a view places proper responsibility on the engineer who is intimately connected with his or her "experiment" and responsible for its safe conduct; moreover, it suggests the following conclusions: For safeguarding the public, precise rules and enforceable sanctions are appropriate components of well-established and regularly reexamined engineering procedures. Little of an experimental nature is probably occurring in such standard activities, and the type of professional conduct required is most likely very clear. In areas where experimentation is involved more substantially, however, rules must not attempt to cover all possible outcomes of an experiment, nor must they force engineers to adopt rigidly specified courses of action. It is here that regulations should be broad, but written to hold engineers accountable for their decisions. Through their professional societies, engineers should also play an active role in establishing (or changing) enforceable rules as well as in enforcing

them, taking great care to forestall conflicts of interest (see Discussion Topic 5, on the Hydrolevel case).

Industrial Standards

There is one area in which industry usually welcomes greater specificity, and that is in regard to standards. Standards facilitate the interchange of components, they serve as ready-made substitutes for lengthy design specifications, and they decrease production costs.

Standards consist of explicit specifications that, when followed with care, ensure that stated criteria for interchangeability and quality will be attained. Examples range from automobile tire sizes and load ratings, to computer languages. Table 7.1 lists purposes of standards and gives some examples to illustrate those purposes.

Standards are established by companies for in-house use and by professional associations and trade associations for industrywide use. They may also be prescribed as parts of laws and official regulations, for example, as in mandatory standards, which become necessary upon lack of adherence to voluntary standards.

Standards help not only the manufacturers, they also benefit the client and the public. They preserve some competitiveness in industry by reducing overemphasis on name brands and giving the smaller manufacturer a chance to compete. They ensure a measure of quality and thus facilitate more realistic trade-off decisions. International standards are becoming a necessity in European and

Criterion	Purpose	Selected examples
Uniformity of physical properties and functions	Accuracy in measurement, interchangeability, ease of handling	Standards of weights, screw dimensions, standard time, film size
Safety and reliability	Prevention of injury, death, and loss of income or property	National Electric Code, boiler code, methods of handling toxic wastes
Quality of product	Fair value for price	Plywood grades, lamp life
Quality of personnel and service	Competence in carrying out tasks	Accreditation of schools, professional licenses
Use of accepted procedures	Sound design, ease of communications	Drawing symbols, test proecdures
Separability	Freedom from interference	Highway lane markings, radio frequency bands
Quality procedures approved by ISO, the International Standards Organization	Assurance of product acceptance in member countries	Quality of products, work, certificates, and degrees

Table 7.1 Types of Standards

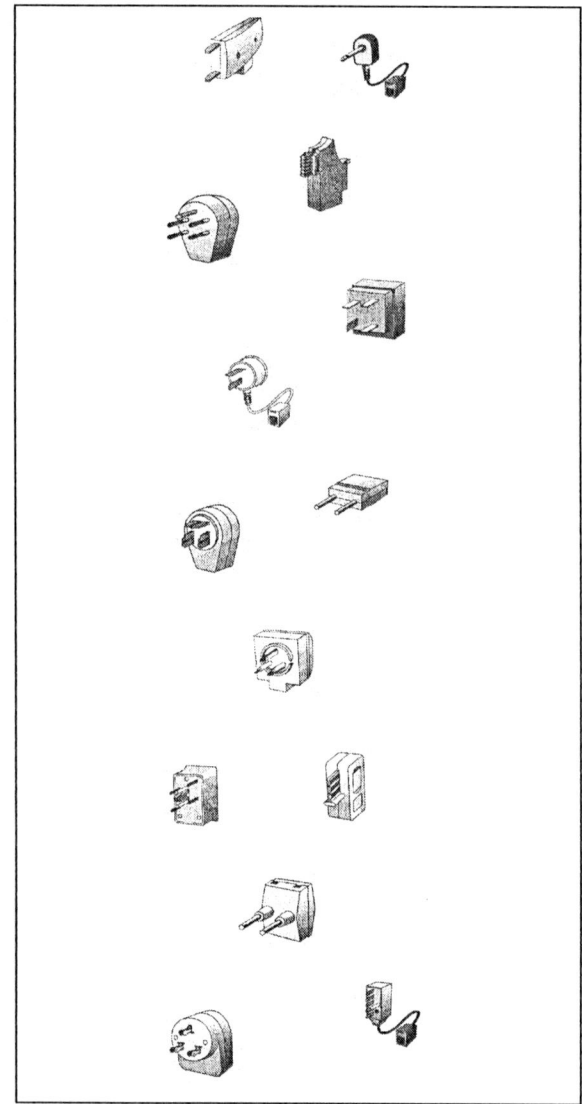

Figure 7-1   A selection of modular telephone adaptors offered by Magellan's.

"Wouldn't it be nice if they could agree on a common telephone jack?"

© Magellan's. A small section from 35 types shown in Magellan's 1999 catalogs. Reprinted with permission.

innovation. The move to performance standards, which in the case of a floor may specify only the required load-bearing capacity, has alleviated that problem somewhat. But other difficulties can arise when special interests (for example, manufacturers, trade unions, exporters and importers) manage to impose unnecessary provisions on standards, or remove important provisions from them, to secure their own narrow self-interest. Requiring metal conduits for home wiring is one example of this problem. Modern conductor coverings have eliminated the need for metal conduit in numerous applications, but many localities still require it. Its use sells more conduit and employs more electricians for installation.

There are standards nowadays for practically everything, it seems, and consequently we often assume that stricter regulation exists than may actually be the case. The public tends to trust the National Electrical Code in all matters of power distribution and wiring, but how many people realize that this code, issued by the National Fire Protection Association, is primarily oriented toward fire hazards? Only recently have its provisions against electric shock begun to be strengthened. Few consumers know that an Underwriter Laboratories seal prominently affixed to the cord of an electrical appliance may pertain only to the cord and not to the rest of the device. In a similar vein, a patent notation inscribed on the handle of a product may refer just to the handle, and then possibly only to the design of the handle's appearance.

Sometimes standards are thought to apply when in actuality there is no standard at all. An example can be found in the widely varying worth and quality of academic degrees—doctorates are even available from mail-order houses. Product appearances can be misleading in this respect. Years ago, when competing foreign firms were attempting to corner the South American market for electrical fixtures and appliances, one manufacturing company had a shrewd idea. It equipped its light bulbs with extra-long bases and threads. These would fit into the competitors' shorter lamp sockets and its own deep sockets. But the competitors' bulbs would not fit into the deeper sockets of its own fixtures (see Figure 7.2). Yet so far as the unsuspecting consumer was concerned, all the light bulbs and sockets continued to look alike.

During the introduction of novel products, there is frequently a period during which the consumer is at a disadvantage, not knowing which type or size of magnetic recording tape, word processing program, or camera lens mount will win in the long run and thus make a recently purchased product prematurely obsolete or nonrepairable. Sometimes a particular design stays in the

world trade. An interesting approach has been adopted by the International Standards Organization (ISO) that replaces the detailed national specifications for a plethora of products with statements of procedures a manufacturer guarantees to carry out in assuring quality products.

Standards have been a hindrance at times. For many years, they were mostly descriptive, specifying, for instance, how many joists of what size should support a given type of floor. Clearly such standards tended to stifle

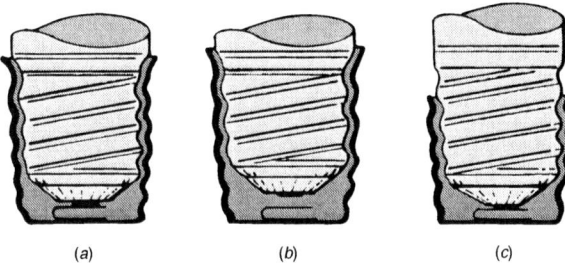

Figure 7.2
The light bulb story. (a) Long base, deep socket: firm contact. (b) Short base, deep socket: no contact. (c) Long base, shallow socket: firm contact.

front long enough until it becomes the standard, as happened to Hayes modems and their command structure. More recently though, a joint effort by otherwise competing photographic film manufacturers resulted in an agreed-upon standard before they introduced the Advanced Photo System (APS) with new standards for film cassettes and processing.

Discussion Topics

1. A common excuse for carrying out a morally questionable project is, "If I don't do it, somebody else will." This rationale may be tempting for engineers who typically work in situations where someone else might be ready to replace them on a project. Do you view it as a legitimate excuse for engaging in projects that might be unethical? (In your answer, comment on the concept of responsible conduct developed in this section.)

2. Another commonly used phrase, "I only work here," implies that one is not personally accountable for the company rules since one does not make them. It also suggests that one wishes to restrict one's area of responsibility within tight bounds as defined by those rules. In light of the discussion in this section, respond to the potential implications of this phrase and the attitude represented by it when exhibited by engineers.

3. Threats to a sense of personal responsibility are neither unique to nor more acute for engineers than they are for others involved with engineering and its results. The reason is that, in general, public accountability also tends to lessen as professional roles become narrowly differentiated. With this in mind, critique each of the remarks made in the following dialog. Is the remark true, or partially true? What needs to be added to make it accurate?

ENGINEER: My responsibility is to receive directives and to create products within specifications set by others. The decision about what products to make and their general specifications are economic in nature and made by management.

SCIENTIST: My responsibility is to gain knowledge. How the knowledge is applied is an economic decision made by management, or else a political decision made by elected representatives in government.

MANAGER: My responsibility is solely to make profits for stockholders.

STOCKHOLDER: I invest my money for the purpose of making a profit. It is up to managers to make decisions about the directions of technological development.

CONSUMER: My responsibility is to my family. Government should make sure corporations do not harm me with dangerous products, harmful side effects of technology, or dishonest claims.

GOVERNMENT REGULATOR: By current reckoning, government has strangled the economy through overregulation of business. Accordingly, at present on my job, especially given decreasing budget allotments, I must back off from the idea that business should be policed and urge corporations to assume greater public responsibility.

4. In 1975, Hydrolevel Corporation brought suit against the American Society of Mechanical Engineers (ASME), charging that two ASME volunteers, acting as agents of ASME, had conspired to interpret a section of ASME's Boiler and Pressure Vessel Code in such a manner that Hydrolevel's low-water fuel cutoff for boilers could not compete with the devices built by the employers of the two volunteers. On May 17, 1982, the Supreme Court upheld the lower courts that had found ASME guilty of violating antitrust provisions and had opened the way for awarding treble damages. (A U.S. district court's award of $7.5 million had been found excessive by the court of appeals and later the award was set at $4.74 million, including legal costs.)

Writing on behalf of the six-to-three majority, Justice Harry A. Blackmun said:

When ASME's agents act in its name, they are able to affect the lives of large numbers of people and the competitive fortunes of businesses throughout the country. By holding ASME liable under the antitrust laws for the antitrust violations of its agents committed with apparent authority, we recognize the important role of ASME and its agents in the economy, and we help to ensure that standard-setting organizations will act with care when they permit their agents to speak for them.

Acquaint yourself with the particulars of this case and discuss it as an illustration of the possible misuses of standards.[18]

5. Mismatched bumpers: Ought there to be a law? What happens when a passenger car rear-ends a truck or a sports utility vehicle (SUV)? The

bumpers usually ride at different heights, so even modest collisions can result in major repair bills. (At high speed, with the front of the car nose down upon braking, people in convertibles have been decapitated upon contact devoid of protection by bumpers.) The men and women at Volvo recognized the problem long ago—we have observed that their trucks usually have low bumpers front and rear. Discuss how other companies building and selling trucks and high-riding vehicles can be induced to follow Volvo's example. Should older vehicles be retrofitted with lower bumpers or guards? Must we have yet another law because otherwise industry will not act on its own?

6. If someone were to ask you to describe codes of ethics for engineers in terms of the three columns provided in Table 7.1, what entries would you provide?

[18] Stephen H. Unger, *Controlling Technology: Ethics and the Responsible Engineer,* 2nd ed. (New York: John Wiley & Sons, 1994), pp. 210–15. See also Paula Wells, Hardy Jones, and Michael Davis, *Conflicts of Interest in Engineering* (Dubuque, IA: Kendall/Hunt, 1986).

Chapter

Workplace Responsibilities and Rights

8

Data General Corporation grew spectacularly during its first decade of operation, quickly becoming a Fortune 500 company that was ranked third in overall sales of small computers. However, it began to fall behind the competition and desperately needed a powerful new microcomputer to sustain its share of the market. The development of that computer is chronicled by Tracy Kidder in his Pulitzer Prize–winning book *The Soul of a New Machine*.

Tom West, one of Data General's most trusted engineers, convinced management that he could build the new computer within one year—an unprecedented time for a project of its importance. West assembled a team of 15 exceptionally motivated though relatively inexperienced young engineers, many of whom were just out of school. Within six months, they designed the central processing unit, and they delivered the complete computer ahead of schedule. Named the Eclipse MV/8000, the computer immediately became a major marketing success.

The remarkable success was possible because the engineers came to identify themselves with the project and the product: "Ninety-eight percent of the thrill comes from knowing that the thing you designed works, and works almost the way you expected it would. If that happens, part of you is in that machine."[1] Yet, the "soul" of the new machine was not any one person. Instead, it was the team of engineers who invested themselves in the product through their personal commitment to work together creatively with colleagues as part of a design group. As might be expected, personality clashes occurred during the sometimes frenzied work schedule, but conflicts were minimized by a commitment to teamwork, collegiality, shared commitment, and identification with the group's project.

This kind of performance understandably ranks high on the list of expectations that employers have of the engineers they employ or of the engineers they engage as con-

sultants. Engineers in turn should see top performance at a professional level as their main responsibility, followed by others such as maintaining confidentiality and avoiding conflicts of interest. But engineers must also be given the opportunity to perform responsibly, and this means that their professional rights must be observed.

In this chapter we first address the needs to maintain confidentiality and to avoid conflicts of interest as issues related to an engineer's responsibilities as an employee. Next we discuss an engineer's rights as a professional and as an employee. We end with the issue of whistle-blowing that can arise when loyalty to the employer is in conflict with the engineer's personal ethics and the engineer's duty as a citizen.

8.1 | ISSUES OF RESPONSIBILITY, CONFIDENTIALITY, AND CONFLICTS OF INTEREST

Maintaining confidentiality and avoiding harmful conflicts of interest are especially important aspects of teamwork and trustworthiness. Let us begin with confidentiality and then turn to conflicts of interest, in each case seeking a clearer understanding of what is at stake morally.

Confidentiality: Definition

The duty of confidentiality is the duty to keep secret all information deemed desirable to keep secret. Deemed by whom? Basically, any information that the employer or client would like to have kept secret in order to compete effectively against business rivals. Often, this is understood to be any data concerning the company's business or technical processes that are not already public knowledge. While this criterion is somewhat vague, it clearly points to the employer or client as the main source of the decision as to what information is to be treated as confidential.

[1] Tracy Kidder, *The Soul of a New Machine* (New York: Avon Books, 1981), p. 273.

"Keep secret" is a relational expression. It always makes sense to ask, "Secret with respect to whom?" In the case of some government organizations, such as the FBI and CIA, highly elaborate systems for classifying information have been developed that identify which individuals and groups may have access to what information. Within other governmental agencies and private companies, engineers and other employees are usually expected to withhold information labeled "confidential" from unauthorized people both inside and outside the organization.

Several related terms should be distinguished. *Privileged information* literally means "available only on the basis of special privilege," such as the privilege accorded an employee working on a special assignment. *Proprietary information* is information that a company owns or is the proprietor of, and hence is a term carefully defined by property law. A rough synonym for "proprietary information" is *trade secret,* which can be virtually any type of information that has not become public, that an employer has taken steps to keep secret, and that is thereby given limited legal protection in common law (law generated by previous court rulings) that forbids employees from divulging it. *Patents* legally protect specific products from being manufactured and sold by competitors without the express permission of the patent holder. Trade secrets have no such protection, and a corporation may learn about a competitor's trade secrets through legal means—for instance, "reverse engineering," in which an unknown design or process can be traced out by analyzing the final product. But patents do have the drawback of being public and thus allowing competitors an easy means of working around them by finding alternative designs.

Confidentiality and Changing Jobs

The obligation to protect confidential information does not cease when employees change jobs. If it did, it would be impossible to protect such information. Former employees would quickly divulge it to their new employers, or perhaps for a price sell it to competitors of their former employers. Thus, the relationship of trust between employer and employee in regard to confidentiality continues beyond the formal period of employment. Unless the employer gives consent, former employees are barred indefinitely from revealing trade secrets. This provides a clear illustration of the way in which the professional integrity of engineers involves much more than mere loyalty to one's present employer.

Yet thorny problems arise in this area. Many engineers value professional advancement more than long-term ties with any one company and so change jobs frequently. Engineers in research and development are especially likely to have high rates of turnover. They are also the people most likely to be exposed to important new trade secrets. Moreover, when they transfer into new companies, they frequently do the same kind of work as before—precisely the type of situation in which trade secrets of their old companies may have relevance, a fact that could have strongly contributed to their having readily found new employment.

A high-profile case of trade secret violations in recent history was settled in January 1997 without coming to trial when VW (Volkswagen AG) agreed to pay GM (General Motors Corp. and GM's German subsidiary Adam Opel) $100 million in cash and to buy $1 billion in parts from GM over the next seven years. Why? Because in March 1993, Jose Ignacio Lopez, GM's highly effective manufacturing expert, left GM to join VW (a fierce competitor in Europe) and took with him not only three colleagues and know-how, but also copies of confidential GM documents.

Instead of examining the Lopez case further, let us consider the legally significant case of Donald Wohlgemuth, a chemical engineer who at one time was manager of B.F. Goodrich's space suit division.[2] Technology for space suits was undergoing rapid development, with several companies competing for government contracts. Dissatisfied with his salary and the research facilities at B. F. Goodrich, Wohlgemuth negotiated a new job with International Latex Corporation as manager of engineering for industrial products. International Latex had just received a large government subcontract for developing the Apollo astronauts' space suits, and that was one of the programs Wohlgemuth would manage.

The confidentiality obligation required that Wohlgemuth not reveal any trade secrets of Goodrich to his new employer. But this was easier said than done. Of course, it is possible for employees in his situation to refrain from explicitly stating processes, formulas, and material specifications. Yet in exercising their general skills and knowledge, it is virtually inevitable that some unintended "leaks" will occur. An engineer's knowledge base generates an intuitive sense of what designs will or will not work, and trade secrets form part of this knowledge base. To fully protect the secrets of an old employer on a new job would thus virtually require that part of the engineer's brain be removed.

Is it perhaps unethical, then, for employees to change jobs in cases where unintentional revelations of confidential

---

[2] Michael S. Baram, "Trade Secrets: What Price Loyalty?" *Harvard Business Review,* November–December 1968. Reprinted in Deborah G. Johnson, ed., *Ethical Issues in Engineering* (Englewood Cliffs, NJ: Prentice Hall, 1991), pp. 279–90.

information are a possibility? Some companies have contended that it is. Goodrich, for example, charged Wohlgemuth with being unethical in taking the job with International Latex. Goodrich also went to court seeking a restraining order to prevent him from working for International Latex or any other company that developed space suits. The Ohio Court of Appeals refused to issue such an order, although it did issue an injunction prohibiting Wohlgemuth from revealing any Goodrich trade secrets. Their reasoning was that while Goodrich had a right to have trade secrets kept confidential, it had to be balanced against Wohlgemuth's personal right to seek career advancement. And this would seem to be the correct moral verdict as well.

Confidentiality and Management Policies

What might be done to recognize the legitimate personal interests and rights of engineers and other employees while also recognizing the rights of employers in this area?[3] One approach is to use employment contracts that place special restrictions on future employment. Traditionally, those restrictions have centered on geographical location of future employers, length of time after leaving the present employer before one can engage in certain kinds of work, and the type of work it is permissible to do for future employers. Thus, Goodrich might have required as a condition of employment that Wohlgemuth sign an agreement that if he sought work elsewhere, he would not work on space suit projects for a competitor in the United States for five years after leaving Goodrich.

Yet such contracts are hardly agreements between equals, and they threaten the right of individuals to pursue their careers freely. For this reason, the courts have tended not to recognize such contracts as binding, although they do uphold contractual agreements forbidding disclosure of trade secrets.

A different type of employment contract is perhaps not so threatening to employee rights in that it offers positive benefits in exchange for the restrictions it places on future employment. Consider a company that normally does not have a portable pension plan. It might offer such a plan to an engineer in exchange for an agreement not to work for a competitor on certain kinds of projects for a certain number of years after leaving the company. Or another clause might offer an employee a special post-employment annual consulting fee for several years on the condition that he or she not work for a direct competitor during that period.

Other tactics aside from employment contract provisions have been attempted by various companies. One is to place tighter controls on the internal flow of information by restricting access to trade secrets except where absolutely essential. The drawback to this approach is that it may create an atmosphere of distrust in the workplace. It might also stifle creativity by lessening the knowledge base of engineers involved in research and development.

One potential solution is for employers to help generate a sense of professional responsibility among their staff that reaches beyond merely obeying the directives of current employers. Engineers can then develop a real sensitivity to the moral conflicts to which they may be exposed by making certain job changes within the company. They can arrive at a greater appreciation of why trade secrets are important in a competitive system and learn to take the steps necessary to protect them. In this way, professional concerns and employee loyalty can become intertwined and reinforce each other.

Confidentiality: Justification

Upon what moral basis does the confidentiality obligation rest, with its wide scope and obvious importance? The primary justification is to respect the autonomy (freedom, self-determination) of individuals and corporations and to recognize their legitimate control over some private information concerning themselves.[4] Without that control, they could not maintain their privacy and protect their self-interest insofar as it involves privacy. Just as patients should be allowed to maintain substantial control over personal information, so employers should have some control over the private information about their companies. All the major ethical theories recognize the importance of autonomy, whether it is understood in terms of rights to autonomy, duties to respect autonomy, the utility of protecting autonomy, or the virtue of respect for others.

Additional justifications include trustworthiness: Once practices of maintaining confidentiality are established socially, trust and trustworthiness can grow. Thus, when clients go to attorneys or tax accountants, they expect them to maintain confidentiality, and the professional indicates that confidentially will be maintained. Similarly, employees often make promises (in the form of signing contracts) not to divulge certain information considered sensitive by the employer.

---

[3] Ibid., pp. 285–90.

[4] Sissela Bok, *Secrets* (New York: Pantheon Books, 1982), pp. 116–35.

In addition, there are public benefits in recognizing confidentiality relationships within professional contexts. For example, if patients are to have the best chances of being cured, they must feel completely free to reveal the most personal information about themselves to physicians, and that requires trust that the physician will not divulge private information. Likewise, the economic benefits of competitiveness within a free market are promoted when companies can maintain some degree of confidentiality concerning their products. Developing new products often requires investing large resources in acquiring new knowledge. The motivation to make those investments might diminish if that knowledge were immediately dispersed to competitors who could then quickly make better products at lesser cost, since they did not have to make comparable investments in research and development.

It must not be overlooked that confidentiality has its limits, particularly when it is invoked to hide misdeeds. Investigations into a wide variety of white collar crimes covered up by management in industry or public agencies have been thwarted by invoking confidentiality or false claims of secrecy based on national interest.

Conflicts of Interest: Definition and Examples

We turn now to some equally thorny issues concerning conflicts of interest. Professional conflicts of interest are situations where professionals have an interest that, if pursued, might keep them from meeting their obligations to their employers or clients. Sometimes such an interest involves serving in some other professional role, say, as a consultant for a competitor's company. Other times it is a more personal interest, such as making substantial private investments in a competitor's company.

Concern about conflicts of interest largely centers on their potential to distort good judgment in faithfully serving an employer or client.[5] Exercising good judgment means arriving at beliefs on the basis of expertise and experience, as opposed to merely following simple rules. Thus, we can refine our definition of conflicts of interest by saying that they typically arise when *two* conditions are met: (1) the professional is in a relationship or role that requires exercising good judgment on behalf of the interests of an employer or client and (2) the professional has some additional or side interest that could threaten good judgment in serving the interests of the employer or client—ei-

ther the good judgment of that professional or the judgment of a typical professional in that situation. Why the reference to "a typical professional"? There might be conclusive evidence that the actual professionals involved would never allow a side interest to affect their judgment, yet they are still held to be in a conflict of interest.

"Conflict of interest" and "conflicting interests" are not synonyms.[6] A student, for example, may have interests in excelling on four final exams. She knows, however, that there is time to study adequately for only three of them, and so she must choose which interest not to pursue. In this case, "conflicting interests" means a person has two or more desires that cannot all be satisfied given the circumstances. But there is no suggestion that it is morally wrong to try pursuing them all. By contrast, in professional conflicts of interest, it is often physically or economically possible to pursue all of the conflicting interests, but doing so could be morally problematic.

Because of the great variety of possible outside interests, conflicts of interest can arise in innumerable ways, and with many degrees of subtlety. We will sample only a few of the more common situations.

Gifts, Bribes, and Kickbacks   A bribe is a substantial amount of money or goods offered beyond a stated business contract with the aim of winning an advantage in gaining or keeping the contract. "Substantial" is a vague term, but it alludes to amounts, beyond acceptable gratuities, that are sufficient to distort the judgment of a typical person. Typically, though not always, bribes are made in secret. Gifts are not bribes as long as they are small gratuities offered in the normal conduct of business, where a small gift used to be jokingly referred to as having the value of something you could "eat, drink, or smoke in a day." Prearranged payments made by contractors to companies or their representatives in exchange for contracts actually granted are called "kickbacks." When suggested by the granting party to the party bidding on the contract, the latter often defends its participation in such an arrangement as having been subjected to "extortion."

A gift is not a bribe if you can eat, drink, or smoke it in a day.

— Old-timers' saying

---

[5] Michael Davis, "Conflict of Interest," *Business and Professional Ethics Journal* 1 (Summer 1982), pp. 17–27; Paula Wells, Hardy Jones, and Michael Davis, *Conflicts of Interest in Engineering* (Dubuque, IA: Kendall/Hunt, 1986).

[6] Joseph Margolis, "Conflict of Interest and Conflicting Interests," in *Ethical Theory and Business,* ed. T. Beauchamp and N. Bowie (Englewood Cliffs, NJ: Prentice Hall, 1979), p. 361.

Often, companies give gifts to selected employees of government agencies or partners in trade. Many such gifts are unobjectionable, some are intended as bribes, and still others create conflicts of interest that do not, strictly speaking, involve bribes. What are the differences? In theory, these distinctions may seem clear, but in practice they become blurry. Bribes are illegal or immoral because they are substantial enough to threaten fairness in competitive situations, while gratuities are of smaller amounts. Some gratuities play a legitimate role in the normal conduct of business, while others can bias judgment like a bribe does. Much depends on the context, and there are numerous gray areas, which is why companies often develop elaborate guidelines for their employees.

Texas Instruments, for example, not only discusses gifts in its policy manual but it also makes available detailed brochures illustrating acceptable and unacceptable gifts within particular contexts. One context is doing business with the U.S. Department of Defense, whose officials are prohibited by federal law from accepting "anything of value" that bears in any way on official government business. Does that mean a Texas Instrument official cannot buy a moderately priced dinner for a Department of Defense official who is visiting? Yes. Does it mean the Texas Instruments employee cannot take the government official to the airport in the company's courtesy van? Maybe. Transportation to the airport may have a substantial value, as does company time. Texas Instruments allows its employees to offer the government official a ride to the airport if "1. The use of commercial transportation is impractical, and 2. Refusing your offer would interfere significantly with their [i.e., the officials'] performance of official duties."[7] Such serpentine policies may seem silly, but in fact they represent good-faith efforts to avoid even the appearance of a conflict of interest in situations of intense public scrutiny and cynicism.

What about more routine business contexts? Is it all right to accept the occasional luncheon paid for by vendors giving sales presentations, or a gift one believes is given in friendship rather than for influence? The guidelines for use with the principles of ethics of ASCE (Sec. 4c) or ASME (Sec. 4e) recommend a hard line on such gratuities: "Engineers shall not solicit nor accept gratuities, directly or indirectly, from contractors, their agents, or other parties dealing with their clients or employers in connection with work for which they are responsible." Many employers would consider this recommendation ex-

treme. Company policies generally ban any gratuities that have more than nominal value or exceed widely and openly that accepted as normal business practice. An additional rule of thumb is: "If the offer or acceptance of a particular gift could have embarrassing consequences for your company if made public, then do not accept the gift."

Interests in Other Companies   Some conflicts of interest consist in having an interest in a competitor's or a subcontractor's business. One blatant example is actually working for the competitor or subcontractor as an employee or consultant. Another example is partial ownership or substantial stockholdings in the competitor's business. Does holding a few shares of stock in a company with which one has occasional dealings constitute a conflict of interest? Usually not, but as the number of shares of stock increases, the issue becomes blurry. Again, is there a conflict of interest if one's spouse works for a subcontractor to one's company? Usually not, but a conflict of interest arises if one's job involves granting contracts to that subcontractor.

Should there be a general prohibition on *moonlighting,* that is, working in one's spare time for another company? That would violate the rights to pursue one's legitimate self-interest. Moonlighting usually creates conflicts of interest only in special circumstances, such as working for competitors, suppliers, or customers. Even then, in rare situations, an employer sometimes gives permission for exceptions, for example, when the experience gained would greatly promote business interests. A special kind of conflict of interest arises, however, when moonlighting leaves one exhausted and thereby harms job performance.[8]

Insider Information   An especially sensitive conflict of interest consists in using "inside" information to gain an advantage or set up a business opportunity for oneself, one's family, or one's friends. The information might concern one's own company or another company with which one does business. For example, engineers might tell their friends about the impending announcement of a revolutionary invention their company has been perfecting, or of their company's plans for a merger that will greatly improve the worth of another company's stock. In doing so, they give those friends an edge on an investment promising high returns. Owning stock in the company for which one works is of course not objectionable, and this is often

[7] *Cornerstone* 2, TI Ethics Office, Texas Instruments, Richardson, Texas, 1989, p. 2.

[8] George L. Reed, "Moonlighting and Professional Responsibility," *Journal of Professional Activities: Proceedings of the American Society of Civil Engineers* 96 (September 1970); pp. 19–23.

encouraged by employers. But such ownership should be based on the same information available to the general public.

While conflict of interest occurs in many professional areas, it may come as a surprise that it also occurs in academic settings. Professors are frequently hired as consultants by companies that need the academic's expertise, and many professors in the sciences and engineering establish their own companies or form partnerships to develop commercially what they have learned through their research at the university. Here are two examples of the problems that can arise.

Experts at public as well as private universities turned down requests by the office of the Attorney General of California to testify for the state in its damage suit against Union Oil and others who were drilling offshore at Santa Barbara when the massive oil leak occurred in 1969. State officials believe that the experts (among them engineering professors) refused to offer expert testimony because they were afraid of losing industry grants and consulting arrangements if they were to testify unfavorably in the eyes of the oil industry.

A professor of electrical engineering at a west coast university was found to have used $144,000 in grant funds to purchase electronic equipment from a company he owned in part. He had not revealed his ownership to the university; he had priced the equipment much higher than market value, and some of the purchased items were never received. The Supplier Information Form and Sole Source Justification Statements had been submitted as required, but with falsified content. In addition, the professor had hired a brother and two sisters for several years, concealing their relationship to him in violation of anti-nepotism rules and paying them for research work they did not perform. All told he had defrauded the university of at least $500,000 in research funds. Needless to say, the professor lost his university position and had to stand trial in civil court when an internal audit and subsequent hearings revealed these irregularities.

These cases indicate that conflicts of interest are not restricted to industry. The academic biotechnology field, with its many professor-entrepreneurs, is particularly vulnerable in this regard.

Moral Status of Conflicts of Interest

What is wrong with employees having conflicts of interest? Most of the answer is obvious from our stated definition: Employee conflicts of interest occur when employees have interests that if pursued could keep them from meeting their obligations to serve the interests of the employer or client for whom they work. Such conflicts of interest should be avoided because they threaten to prevent one from fully meeting those obligations.

More than this needs to be said, however. Why should mere threats of possible harm always be condemned? Suppose that substantial good might sometimes result from pursuing a conflict of interest?

In fact, it is not always unethical to pursue conflicts of interest. In practice, some conflicts are thought to be unavoidable, or even acceptable. One illustration of this is that the government allows employees of aircraft manufacturers to serve as government inspectors for the Federal Aviation Agency (FAA). The FAA is charged with regulating airplane manufacturers and making objective safety and quality inspections of the airplanes they build. Naturally, the two roles of (1) government inspector and (2) employee of the manufacturer being inspected could bias judgments. Yet with careful screening of inspectors, the likelihood of such bias is said to be outweighed by the practical necessities of airplane inspection. The options would be to greatly increase the number of nonindustry government workers (at great expense to taxpayers) or to do without government inspection altogether (putting public safety at risk).

Even when conflicts of interest are unavoidable or reasonable, employees are still obligated to inform their employers and obtain approval. This suggests a fuller answer to why conflicts of interest are generally prohibited: (1) The professional obligation to employers is very important in that it overrides in the vast majority of cases any appeal to self-interest on the job and (2) the professional obligation to employers is easily threatened by self-interest (given human nature) in a way that warrants especially strong safeguards to ensure that it is fulfilled by employees.

Many conflicts of interest violate trust, in addition to undermining specific obligations. Employed professionals are in fiduciary (trust) relationships with their employers and clients. Allowing side interests to distort one's judgment violates that trust. And additional types of harm can arise as well. Many conflicts of interest are especially objectionable in business affairs precisely because they pose risks to free competition. In particular, bribes and large gifts are objectionable because they lead to awarding contracts for reasons other than the best work for the best price.

As a final point, we should note that even the appearance of conflicts of interest, especially appearances of seeking a personal profit at the expense of one's employer, is considered unethical since the appearance of wrongdoing can harm a corporation as much as any actual bias that might result from such practices.

Discussion Topics

1. Consider the following example:

   Who Owns Your Knowledge? Ken is a process engineer for Stardust Chemical Corp., and he has signed a secrecy agreement with the firm that prohibits his divulging information that the company considers proprietary.

   Stardust has developed an adaptation of a standard piece of equipment that makes it highly efficient for cooling a viscous plastics slurry. (Stardust decides not to patent the idea but to keep it as a trade secret.) Eventually, Ken leaves Stardust and goes to work for a candy-processing company that is not in any way in competition. He soon realizes that a modification similar to Stardust's trade secret could be applied to a different machine used for cooling fudge, and at once has the change made.[9]

   Has Ken acted unethically?

2. American Potash and Chemical Corporation advertised for a chemical engineer having industrial experience with titanium oxide. It succeeded in hiring an engineer who had formerly supervised E. I. Du Pont de Nemours and Company's production of titanium oxide. Du Pont went to court and succeeded in obtaining an injunction prohibiting the engineer from working on American Potash's titanium oxide projects. The reason given for the injunction was that it would be inevitable that the engineer would disclose some of du Pont's trade secrets.[10] Defend your view as to whether the court injunction was morally warranted or not.

3. Consider the following case:

   Facts: Engineer Doe is employed on a full-time basis by a radio broadcast equipment manufacturer as a sales representative. In addition, Doe performs consulting engineering services to organizations in the radio broadcast field, including analysis of their technical problems and, when required, recommendation of certain radio broadcast equipment as may be needed. Doe's engineering reports to his clients are prepared in form for filing with the appropriate governmental body having jurisdiction over radio broadcast facilities. In some cases Doe's engineering reports recommend the use of broadcast equipment manufactured by his employer.

   Question: May Doe ethically provide consulting services as described?[11]

4. Consider the following case:

   Scott Bennett is the engineer assigned to deal with vendors who supply needed parts to the Upscale Company. Larry Newman, sales representative from one of Upscale's regular vendors, plays in the same golf league as Scott. One evening they go off in the same foursome. Sometime during the round Scott mentions that he is really looking forward to vacationing in Florida next month. Larry says his uncle owns a condo in Florida that he rents out during the months he and his family are up north. Larry offers to see if the condo is available next month—assuring Scott that the rental cost would be quite moderate.

   What should Scott say?[12]

   Does your answer turn on whether Scott's company policy indicates a clear answer to this question?

5. Junior colleges and even universities increasingly hire part-time lecturers to teach classes when they lack the funds to employ tenure-truck faculty, or when they think they can get more for less. But they usually will not pay a lecturer enough per course, nor engage any one lecturer for enough courses at a time, to allow a decent living. Nor will such part-time lecturers receive the usual health and retirement benefits. As a result, many lecturers become "freeway flyers" as they rush from one college to another to add to their meager income from each. Schools of engineering may hire a lecturer on a temporary basis when such a person can give instruction in a specialty field not covered by the regular faculty, or while a professorship is vacant. What do you see as advantages and disadvantages of hiring part-time lecturers? Are there dangers of *conflicting interests* when a lecturer's time is taken up by too many commitments at the same time? Could there also be *conflicts of interest*? Do these situations involve moonlighting?

---

[9] Philip M. Kohn and Roy V. Hughson, "Perplexing Problems in Engineering Ethics," *Chemical Engineering* 87 (May 5, 1980), p. 102. Quotations in text used with permission of McGraw-Hill Book Co.

[10] Charles M. Carter, "Trade Secrets and the Technical Man," *IEEE Spectrum* 6 (February 1969), p. 54.

[11] *NSPE Opinions of the Board of Ethical Review*, Case 75.10, National Society of Professional Engineers, Washington, DC, web site: www.nspe.org/eh-home.asp.

[12] "The Condo," in *Teaching Engineering Ethics: A Case Study Approach*, ed. Michael S. Pritchard (Kalamazoo: Center for the Study of Ethics in Society, Western Michigan University, 1993).

8.2 | RIGHTS OF ENGINEERS

Engineers have several types of moral rights, which fall into the sometimes overlapping categories of human, employee, contractual, and professional rights. As *humans,* engineers have fundamental rights to live and freely pursue their legitimate interests, which implies, for example, rights not to be unfairly discriminated against in employment on the basis of sex, race, or age. As *employees,* engineers have special rights, including the right to receive one's salary in return for performing one's duties and the right to engage in the nonwork political activities of one's choosing without reprisal or coercion from employers. As *professionals,* engineers have special rights that arise from their professional role and the obligations it involves. We begin with professional rights, most of which can be viewed as aspects of a fundamental right of professional conscience. We now move to a discussion of professional rights, followed by employee rights.

Professional Rights

Basic Right of Professional Conscience
The right of professional conscience is the moral right to exercise professional judgment in pursuing professional responsibilities. Pursuing those responsibilities involves exercising both technical judgment and reasoned moral convictions. This right has limits, of course, and must be balanced against responsibilities to employers and colleagues of the sort discussed earlier.

If the duties of engineers were so clear that it was obvious to every sane person what was morally proper in every situation, there would be little point in speaking of "conscience" in specifying this basic right. Instead, we could simply say it is the right to do what everyone agrees is obligatory for the professional engineer to do. But engineering, like other professions, calls for morally complex decisions. It requires autonomous moral judgment in attempting to uncover the most morally reasonable courses of action, and the correct courses of action are not always obvious.

As with most moral rights, the basic professional right is an entitlement giving one the moral authority to act without interference from others. It is a "liberty right" that places an obligation on others not to interfere with its proper exercise. Yet, occasionally, special resources may be required by engineers seeking to exercise the right of professional conscience in the course of meeting their professional obligations. For example, conducting an adequate safety inspection may require that special equip-

ment be made available by employers. Or, more generally, in order to feel comfortable about making certain kinds of decisions on a project, the engineers involved may need an environment conducive to trust and support, which management may be obligated to help create and sustain. In this way, the basic right is also in some respects a "positive right" placing on others an obligation to do more than merely not interfere.

Character counts. Ethics is not for whimps.

— Michael Josephson

The right of professional conscience implies more specific rights, corresponding to specific professional obligations. In the next part of this chapter, we discuss the right to whistle blow in some situations where the public good is severely threatened. Here we cite two further examples: the right of conscientious refusal and the right to recognition.

Right of Conscientious Refusal
The right of conscientious refusal is the right to refuse to engage in unethical behavior, and to refuse to do so solely because one views it as unethical. This is a kind of second-order right. It arises because other rights to honor moral obligations within the authority-based relationships of employment sometimes come into conflict.

There are two situations to be considered: (1) where there is widely shared agreement in the profession as to whether an act is unethical and (2) where there is room for disagreement among reasonable people over whether an act is unethical.

It seems clear enough that engineers and other professionals have a moral right to refuse to participate in activities that are illegal and uncontroversially unethical (for example, forging documents, altering test results, lying, giving or taking bribes, or padding payrolls). And coercing employees into acting by means of threats (to their jobs) plainly constitutes a violation of this right of theirs.

The troublesome cases concern situations where there is no shared agreement about whether a project or procedure is unethical. Do engineers have any rights to exercise their personal consciences in these more cloudy areas? Just as pro-life physicians and nurses have a right not to participate in abortions, engineers should be recognized as having a *limited* right to turn down assignments that violate their personal consciences in matters of great importance, such as threats to human life, even where there is room for moral disagreement among reasonable people about the situation in question. We emphasize the word "limited" because the right is contingent on the organization's ability to reassign the engineer to alternative projects without serious economic hardship to itself. The right of professional conscience does not extend to the right to be paid for not working.

Right to Recognition Engineers have a right to professional recognition for their work and accomplishments. Part of this involves fair monetary remuneration and part, nonmonetary forms of recognition. The right to recognition, and especially fair remuneration, may seem to be purely a matter of self-interest rather than morality, but it is both. Without a fair remuneration, engineers cannot concentrate their energies where they properly belong—on carrying out the immediate duties of their jobs and on maintaining up-to-date skills through formal and informal continuing education. Their time will be taken up by money worries, or even by moonlighting in order to maintain a decent standard of living.

The right to reasonable remuneration is clear enough to serve as a moral basis for arguments against corporations that make excessive profits while their employees are poorly paid. It can also serve as the basis for criticizing the unfairness of patent arrangements that fail to give more than nominal rewards to the creative engineers who make the discoveries leading to the patents. If a patent leads to millions of dollars of revenue for a company, it is unfair to give the discoverer no more than a nominal bonus and a thank-you letter.

But the right to professional recognition is not sufficiently precise to pinpoint just what a reasonable salary is or what a fair remuneration for patent discoveries is. Such detailed matters must be worked out cooperatively between employers and employees, for they depend on both the resources of a company and the bargaining position of engineers. Professional societies can be of help by providing general guidelines.

Foundation of Professional Rights There are two general ways to justify the basic right of professional conscience. One is to proceed piecemeal by reiterating the justifications given for the specific professional *duties*. Whatever justification there is for the specific duties will also provide justification for allowing engineers the *right* to pursue those duties. Fulfilling duties, in turn, requires the exercise of moral reflection and conscience, rather than rote application of simplistic rules. Hence, the justification of each duty ultimately yields a justification of the right of conscience with respect to that duty.

The second way is to justify the right of professional conscience, which involves grounding it more directly in the ethical theories. Thus, duty ethics regards professional rights as implied by general duties to respect persons, and rule-utilitarianism would accent the public good of allowing engineers to pursue their professional duties. Rights ethics would justify the right of professional conscience by reference to the rights of the public not to be harmed and the rights to be warned of dangers from the "social experiments" of technological innovation.

Employee Rights

Employee rights are any rights, moral or legal, that involve the status of being an employee. They overlap with some professional rights, of the sort just discussed, and they also include institutional rights created by organizational policies or employment agreements, such as the right to be paid the salary specified in one's contract or employment agreement. However, here we will focus on human rights that exist even if unrecognized by specific contract arrangements.

Many of these human rights are discussed more fully in *Freedom Inside the Organization* by David Ewing[13] (for many years editor of *The Harvard Business Review,* 1949–85). Ewing refers to employee rights as the "black hole in American rights." The Bill of Rights in the Constitution was written to apply to government, not to business. But when the Constitution was written, no one envisaged the giant corporations that have emerged in our century. Corporations wield enormous power politically and socially, often in multinational settings, and they operate much as mini-governments and are often comparable in size to those governments the authors of the Constitution had in mind. For instance, American Telephone & Telegraph in the 1970s employed twice the number of people inhabiting the largest of the original 13 colonies when the Constitution was written.

Ewing proposes that large corporations ought to recognize a basic set of employee rights. As examples, we will discuss rights to privacy and to equal opportunity.

Privacy The right to pursue outside activities can be thought of as a right to personal privacy in the sense that it means the right to have a private life off the job. In speaking of the right to privacy here, however, we mean the right to control access to and use of information about oneself. As with the right to outside activities, this

[13] David W. Ewing, *Freedom Inside the Organization* (New York: McGraw-Hill, 1977), pp. 234–35.

right is limited in certain instances by employers' rights, but even then who among employers has access to confidential information is restricted. For example, the personnel division needs medical and life insurance information about employees, but immediate supervisors usually do not.

Consider a few examples of situations in which the functions of employers conflict with the right employees have to privacy:

1. Job applicants at the sales division of an electronics firm are required to take personality tests that include personal questions about alcohol use and sexual conduct. The rationale given for asking those questions is a sociological study showing correlations between sales ability and certain data obtained from answers to the questionnaire. (That study has been criticized by other sociologists.)

2. A supervisor unlocks and searches the desk of an engineer who is away on vacation without the permission of that engineer. The supervisor suspects the engineer of having leaked information about company plans to a competitor and is searching for evidence to prove those suspicions.

3. A large manufacturer of expensive pocket computers has suffered substantial losses from employee theft. It is believed that more than one employee is involved. Without notifying employees, hidden surveillance cameras are installed.

4. A rubber products firm has successfully resisted various attempts by a union to organize its workers. It is always one step ahead of the union's strategies, in part because it monitors the phone calls of employees who are union sympathizers. It also pays selected employees bonuses in exchange for their attending union meetings and reporting on information gathered. It considered, but rejected as imprudent, the possibility of bugging the rest areas where employees were likely to discuss proposals made by union organizers.

We may disagree about which of these examples involve abuse of employer prerogatives. Yet the examples remind us of both the importance of privacy (as discussed in Chapter 2) and how easily rights of privacy are abused. Employers should be viewed as having the same trust relationship with their employees concerning confidentiality that doctors have with their patients and lawyers have with their clients.[14] In all of these cases, personal information is given in trust on the basis of a special professional relationship.

Equal Opportunity: Nondiscrimination   Perhaps nothing is more demeaning than to be discounted because of personal attributes such as one's race, skin color, age, politics, or religious outlook. These aspects of biological makeup and basic conviction lie at the heart of self-identity and self-respect. Such discrimination—that is, morally unjustified treatment of people on arbitrary or irrelevant grounds—is especially pernicious within the work environment, for work is itself fundamental to a person's self-image. Accordingly, human rights to fair and decent treatment at the workplace and in job training are vitally important.

Consider the following examples:

1. An opening arises for a chemical plant manager. Normally such positions are filled by promotions from within the plant. The best qualified person in terms of training and years of experience is an African-American engineer. Management believes, however, that the majority of workers in the plant would be disgruntled by the appointment of a nonwhite manager. They fear lessened employee cooperation and efficiency. They decide to promote and transfer a white engineer from another plant to fill the position.

2. A farm equipment manufacturer has been hit hard by lowered sales caused by a flagging produce economy. Layoffs are inevitable. During several clandestine management meetings, it is decided to use the occasion to "weed out" some of the engineers within 10 years of retirement in order to avoid payments of unvested pension funds.

These examples involve discrimination. They also involve violation of antidiscrimination laws, in particular the Civil Rights Act of 1964: "It shall be an unlawful employment practice for an employer to fail or refuse to hire or to discharge any individual, or otherwise to discriminate against any individual with respect to his compensation, terms, conditions, or privileges of employment, because of such individual's race, color, religion, sex, or national origin" (Title VII, Equal Employment Opportunity). Age discrimination was added in the 1967 Age Discrimination in Employment Act, and discrimination based on disability was forbidden in the 1994 Americans With Disabilities Act.

Equal Opportunity: Sexual Harassment   Beginning in 1991, several events focused national attention on sexual harassment. In October of that year, Anita Hill testified

---

[14] Mordechai Mironi, "The Confidentiality of Personnel Records," *Labor Law Journal* 25 (May 1974), p. 289.

against confirming Supreme Court nominee Clarence Thomas, charging that he made lewd remarks and unwanted sexual provocations to her years earlier when she had worked for him at the Justice Department. Hill was a respected attorney and law professor, and at the time, one-third of Americans were convinced she was telling the truth. The majority of the Senate Hearing Committee sided with Clarence Thomas, however, and he was confirmed as Supreme Court justice amid controversies over what sexual harassment is, how it is to be proven, and how best to prevent it. A series of scandals followed in the military, first the Navy and then the Army, and quickly corporations were caught up in a fundamental social debate about what sexual harassment is. More recently officers of the highest ranks in the U.S. military have been discharged for engaging in sexual relations with wives of subordinates, even when conducted on a consensual basis. President Clinton did not escape censure when found to have been involved in an unseemly sexual liaison with a White House intern, not against her will. Persons in high places must recognize that the aura of importance can attract members of the opposite sex and that exploiting this tendency can damage the working climate and the real business that should be conducted at the workplace.

One definition of sexual harassment is "the unwanted imposition of sexual requirements in the context of a relationship of unequal power."[15] It takes two main forms: *quid pro quo* and hostile work environment.

Quid pro quo includes cases where supervisors require sexual favors as a condition for some employment benefit (a job, promotion, or raise). It can take the form of a sexual threat (of harm) or sexual offer (of a benefit in return for a benefit). *Hostile work environment,* by contrast, is any sexually oriented aspect of the workplace that threatens employees' rights to equal opportunity. It includes unwanted sexual proposals, lewd remarks, sexual leering, posting of nude photos, and inappropriate physical contact.

What is morally objectionable about sexual harassment? Sexual harassment is a particularly invidious form of sex discrimination, involving as it does not only the abuse of gender roles and authority relationships, but the abuse of sexual intimacy itself. Sexual harassment is a display of power and aggression through sexual means. Accordingly, it has appropriately been called "dominance eroticized."[16] Insofar as it involves coercion, sexual ha-

rassment constitutes an infringement of one's autonomy to make free decisions concerning one's body. But whether or not coercion and manipulation are used, it is an assault on the victim's dignity. In abusing sexuality, such harassment degrades people on the basis of a biological and social trait central to their sense of personhood.

Thus, a duty ethicist like Kant would condemn it as violating the duty to treat people with respect, to treat them as having dignity and not merely as means to personal aggrandizement and gratification of one's sexual and power interests. A rights ethicist would see it as a serious violation of the human right to pursue one's work free from the pressures, fears, penalties, and insults that typically accompany sexual harassment. And a utilitarian would emphasize the impact it has on the victim's happiness and self-fulfillment, and on women in general. This also applies to men who experience sexual harassment.

Equal Opportunity: Preferential Treatment Preferential treatment, as we use the expression here, is giving an advantage to a member of a group that in the past was denied equal treatment, in particular, women and minorities. It "reverses" the historical order of preferences. The Supreme Court has largely forbidden the use of explicit numerical quotas for minorities, but it has not forbidden taking into account minority status in all contexts. The *weak form* consists of hiring a woman or a member of a minority over an equally qualified white male. The *strong form,* by contrast, consists of giving preference to women or minorities over better-qualified white males. Can such preference ever be justified? There are compelling arguments on both sides of the issue.[17]

Arguments favoring preferential treatment take three main forms. First, there is an argument based on compensatory justice: Past violations of rights must be compensated. Ideally such compensation should be given to individuals who in the past were denied jobs. But the costs and practical difficulties of determining such discrimination on a case-by-case basis through job-interviewing suggests instead giving preference on the basis of membership in a group that has been disadvantaged in the past. Second, sexism and racism still permeate our society today, and to counterbalance their insidious impact, reverse preferential treatment is warranted in order to ensure equal opportunity for minorities and women. Third, those utilitarians who favor reverse preferential treatment point to its good

[15] Catherine A. MacKinnon, *Sexual Harassment of Working Women* (New Haven, CT: Yale University Press, 1978), pp. 1, 57–82.
[16] Ibid., p. 162.

[17] Steven M. Cahn, ed., *The Affirmative Action Debate* (New York: Routledge, 1995); George E. Curry, ed., *The Affirmative Action Debate* (Reading, MA: Addison-Wesley, 1996).

consequences: integrating women and minorities into the economic and social mainstream (especially in male-dominated professions like engineering), providing role models for minorities that build self-esteem, and strengthening corporate diversity.

Arguments against reverse preferential treatment condemn it as "reverse discrimination." It is a straightforward violation of the rights to equal opportunity of white males and others who are now not given a fair chance to compete on the basis of their qualifications. Granted, past violations of rights may call for compensation, but only compensation to specific individuals who are wronged and only in ways that do not violate the rights of others who did not personally wrong minorities. It is also permissible to provide special funding for educational programs for economically disadvantaged children, but not to use jobs as a compensatory device. Moreover, those utilitarians who are opposed to reverse preferential treatment point to its negative effects: lowering economic productivity by using criteria other than qualifications in hiring, encouraging racism by generating intense resentment generated among white males and their families, encouraging traditional stereotypes that minorities and women cannot make it on their own without special help, and thereby adding to self-doubts of members of these groups.

Various attempts have been made to develop intermediate positions sensitive to all the above arguments for and against strong preferential treatment. For example, one approach rejects blanket preferential treatment of special groups as inherently unjust, but permits reverse preferential treatment within companies that can be shown to have a history of bias against minorities or women. Another approach is to permit weak reverse preferential treatment but to forbid strong forms.

Discussion Topics

1. Present and defend your view as to whether preferential treatment of women and minorities is ever justified.

2. The majority of employers have adopted mandatory random drug testing on their employees, arguing that the enormous damage caused by the pervasive use of drugs in our society carries over into the workplace. Typically, the tests involve taking urine or blood samples, obtained under close observation, thereby raising questions about personal privacy as well as privacy issues about drug usage away from the workplace that is revealed by the tests. Present and defend your view concerning mandatory drug tests at the workplace.

In your answer, take account of the argument set forth by Joseph R. DesJardins and Ronald Duska that, except where safety is a clear and present danger (as in the work of pilots, police, and the military), such tests are unjustified.[18] They contend that employers have a right to the level of performance for which they pay employees, a level typically specified in contracts and job descriptions. When a particular employee fails to meet that level of performance, then employers will take appropriate disciplinary action based on observable behavior. Either way, it is employee performance that is relevant in evaluating employees, not drug usage per se.

3. A company advertises for an engineer to fill a management position. Among the employees the new manager is to supervise is a woman engineer, Ms. X, who was told by her former boss that she would soon be assigned tasks with increased responsibility. The prime candidate for the manager's position is Mr. Y, a recent immigrant from a country known for confining the roles for women. Ms. X was alerted by other women engineers to expect unchallenging, trivial assignments from a supervisor with Mr. Y's background. Is there anything she can and should do? Would it be ethical for her to try to forestall the appointment of Mr. Y?

4. Jim Serra, vice president of engineering, must decide who to recommend for a new director-level position that was formed by merging the product (regulatory) compliance group with the environmental testing group.[19] The top inside candidate is Diane Bryant, senior engineering group manager in charge of the environmental testing group. Bryant is 36, exceptionally intelligent and highly motivated, and a well-respected leader. She is also five months pregnant and is expected to take an eight-week maternity leave two months before the first customer ship deadline (six months away) for a new product. Bryant applies for the job and, in a discussion with Serra, assures him that she will be available at all crucial stages of the project. Your

18 Joseph DesJardins and Ronald Duska, "Drug Testing in Employment," *Business & Professional Ethics Journal* 6 (1987), pp. 3–21.

19 This case is a summary of Cindee Mock and Andrea Bruno, "The Expectant Executive and the Endangered Promotion," *Harvard Business Review*, January–February 1994, pp. 16–18.

colleague, David Moss, who is vice president of product engineering, strongly urges you to find an outside person, insisting that there is no guarantee that Bryant will be available when needed. Much is at stake. A schedule delay could cost several millions of dollars in revenues lost to competitors. At the same time, offending Bryant could lead her and perhaps other valuable engineers whom she supervises to leave the company. What procedure would you recommend in reaching a solution?

5. Engineering societies have generally portrayed participation by engineers in unions and collective bargaining in engineering as unprofessional and disloyal to employers. Thus, the NSPE code of ethics states, "Engineers shall not actively participate in strikes, picket lines, or other collective coercive action" (Sec. III, 1e). Critics reply that such generalized prohibitions reflect the excessive degree to which engineering is still dominated by corporations' interests. Discuss this issue with regard to the following case. Would the approach de-

scribed below provide an effective alternative way of drawing attention to safety problems? What other options might be pursued, and would they still involve "collective coercive action"?

Management at a mining and refinery operation have consistently kept wages below industrywide levels. They have also sacrificed worker safety in order to save costs by not installing special structural reinforcements in the mines, and they have made no effort to control excessive pollution of the work environment. As a result, the operation has reaped larger-than-average profits. Management has been approached both by individuals and by representatives of employee groups about raising wages and taking the steps necessary to ensure worker safety, but to no avail. A nonviolent strike is called and the metallurgical engineers support it for reasons of worker safety and public health.

We need engineers with the courage to speak out when things are not right, and colleagues to support them when the need arises.

— Schinzinger & Martin

---

[20] We adopt the fourth condition from Marcia P. Miceli and Janet P. Near, *Blowing the Whistle: The Organizational and Legal Implications for Companies and Employees* (New York: Lexington Books, 1992), p. 15.

Global Issues 9

An American family purchasing a General Motors Pontiac Le Mans in 1990 probably believed their purchase would help American auto workers far more so than would the purchase of a foreign-made car. According to Labor Secretary Robert Reich's estimates, however, only $4,000 of the $10,000 sticker price would go directly to Americans. Indeed, that $4,000 would not go to American assembly-line workers, but instead mostly to Detroit strategists in higher management, New York bankers and attorneys, insurance workers spread throughout the country, and General Motors shareholders who include both Americans and non-American foreign investors. The remaining $6,000 would be distributed as follows: "about $3,000 goes to South Korea for routine labor and assembly operations, $1,750 to Japan for advanced components (engines, transaxles, and electronics), $750 to West Germany for styling and design components, $250 to Britain for advertising and marketing services, and about $50 to Ireland and Barbados for data processing."[1] A president of GM once remarked that what is good for GM is good for America, but apparently, in this case, it is even better for the rest of the world.

As both workers and consumers, all of us live in an increasingly international marketplace. American and foreign consumers alike will force companies to compete within a global market. Politically, too, our lives are interwoven with those of people around the world. Our existence as a human species depends on sustaining an increasingly fragile environment. As we enter the twenty-first century, our economic well-being, national security, and biological existence are interdependent with other nations in ways that only a soothsayer could have foreseen a century ago.

The word global in this chapter's title refers to both the international context of engineering and the increasingly pervasive social and environmental dimensions of engineers' work. As responsible social experimenters, engineers need to take these dimensions into account in making engineering decisions and career choices. We will explore these dimensions by discussing three topics:

multinational corporations, environmental ethics, and weapons development.

9.1 | MULTINATIONAL CORPORATIONS

Multinational corporations do extensive business in more than one country. In some cases, the operations of corporations are spread so thinly around the world that the location of their official headquarters in any one home country is largely incidental and essentially a matter of historical circumstance. The benefits to U.S. companies of doing business in less economically developed countries are clear: inexpensive labor, availability of natural resources, favorable tax arrangements, and fresh markets for products. The benefits to the participants in developing countries are equally clear: new jobs, jobs with higher pay and greater challenge, transfer of advanced technology, and an array of social benefits from sharing wealth.

Yet moral difficulties arise, along with business and social complications. Who loses jobs at home when manufacturing is taken "offshore"? What does the host country lose in resources, control over its own trade and standards, and political independence? To what extent are "multinationals" influencing and even usurping the roles of national governments as they play off one government against another? (For instance, see Discussion Topic 6.) And what are the moral responsibilities of corporations and individuals operating in less economically developed countries? Here we focus on the last question. Before doing so, we think it helpful to introduce the concepts of technology transfer and appropriate technology.

Technology Transfer and Appropriate Technology

Technology transfer is the process of moving technology to a novel setting and implementing it there.[2] Technology

---

[1] Robert B. Reich, *The Work of Nations* (New York: Vintage Books, 1992), p. 113.

[2] Peter B. Heller, *Technology Transfer and Human Values* (New York: University Press of America, 1985), p. 119.

includes both hardware (machines and installations) and technique (technical, organizational, and managerial skills and procedures). A novel setting is any situation containing at least one new variable relevant to the success or failure of a given technology. The setting may be within a country where the technology is already used elsewhere, or a foreign country, which is our present interest. A variety of agents may conduct the transfer of technology: governments, universities, private volunteer organizations, consulting firms, and multinational corporations.

In most instances, the transfer of technology from a familiar to a new environment is a complex process. The technology being transferred may be one that originally evolved over a period of time and is now being introduced as a ready-made, completely new entity into a different setting. Discerning how the new setting differs from familiar contexts requires the imaginative and cautious vision of "cross-cultural social experimenters."

The expression appropriate technology is widely used, but with a variety of meanings. We use it in a generic sense to refer to identification, transfer, and implementation of the most suitable technology for a new set of conditions. Typically the conditions include social factors that go beyond routine economic and technical engineering constraints. Identifying them requires attention to an array of human values and needs that may influence how a technology affects the novel situation. Thus, "appropriateness may be scrutinized in terms of scale, technical and managerial skills, materials/energy (assured availability of supply at reasonable cost), physical environment (temperature, humidity, atmosphere, salinity, water availability, etc.), capital opportunity costs (to be commensurate with benefits), but especially human values (acceptability of the end-product by the intended users in light of their institutions, traditions, beliefs, taboos, and what they consider the good life)."[3]

Examples include the introduction of agricultural machines and long-distance telephones. A country with many poor farmers can make better immediate use of small, single- or two-wheel tractors that can serve as motorized plows, to pull wagons or to drive pumps, than it can of huge diesel tractors that require collectivized or agribusiness-style farming. On the other hand, the same country may benefit more from the latest in wireless communication technology to spread its telephone service to more people and over long distances than it can from old-fashioned transmission by wire.

Appropriate technology also implies that the technology should contribute to and not distract from sustainable development of the host country by providing for careful stewardship of its natural resources and not degrading the environment beyond its carrying capacity.

Appropriate technology overlaps with, but is not reducible to, intermediate technology, which lies between the most advanced forms available in industrialized countries and comparatively primitive forms in less-developed countries.[4] The British economist E. F. Schumacher argued that intermediate technologies are preferable because the most advanced technologies usually have harmful side effects, such as causing mass migrations from rural areas to cities where corporations tend to locate. These migrations cause overcrowding, and with it poverty, crime, and disease. Far more appropriate, he argued, are smaller-scale technologies replicated throughout a less-developed country, using low capital investment, labor intensiveness to provide needed jobs, local resources where possible, and simpler techniques manageable by the local population given its education facilities.

We mention intermediate technology, and the movement inspired by Schumacher, not to offer a general endorsement (often it has been dramatically beneficial; at other times, not particularly effective), but to emphasize that it is only one conception of appropriate technology. "Appropriate technology" is a generic concept that applies to all attempts to emphasize wider social factors when transferring technologies. As such, it reinforces and amplifies our view of engineering as social experimentation.

With these distinctions in mind, let us turn in some detail to a case study illustrating the complexities of engineering within multinational settings.

Bhopal

Union Carbide in 1984 operated in 37 "host countries" in addition to its "home country," the United States, ranking 35th in size among U.S. corporations. On December 3, 1984, the operators of Union Carbide's plant in Bhopal, India, became alarmed by a leak and overheating in a storage tank. The tank contained methyl isocyanate (MIC), a toxic ingredient used in pesticides. As a concentrated gas, MIC burns any moist part of bodies with which it comes in contact, scalding throats and nasal passages, blinding eyes, and destroying lungs. Within an hour, the leak exploded in a gush that sent 40 tons of deadly gas

[3] Ibid.

[4] E. F. Schumacher, *Small Is Beautiful* (New York: Harper & Row, 1973).

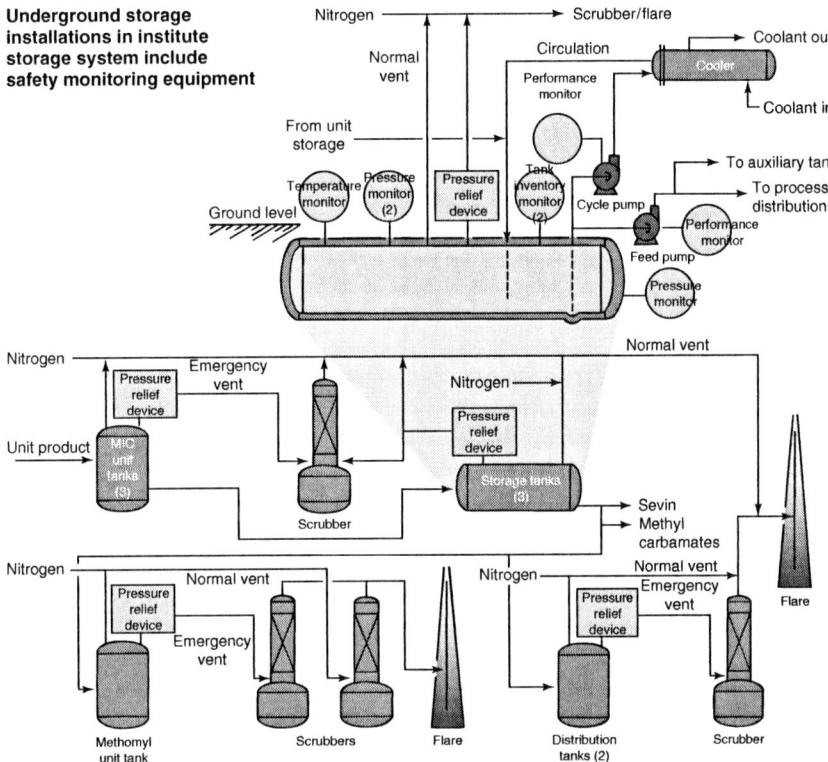

Figure 9.1 Diagram of Bhopal system.
(From Ward Worthy, "Methyl Isocyanate: The Chemistry of a Hazard," *C & EN*, February 11, 1985.)
Reproduced with permission from *Chemical Engineering News*, Feb. 11, 1985, *63* (66), p. 29. Copyright 1985
American Chemical Society.

into the atmosphere.[5] The result was the worst industrial accident in history: 500,000 persons exposed to the gas, 2500 deaths within a few days, 10,000 permanently disabled, 100,000 others injured, and a great loss of livestock. By 10 years later, 12,000 death claims and 870,000 personal injury claims had been submitted, but only $90 million of Union Carbide's $470 million settlement amount had been distributed.

The disaster was caused by a combination of extremely lax safety procedures, gross judgment errors by local plant operators, and possible sabotage with unintended consequences. In retrospect, it is clear that greater sensitivity to social factors was needed in transferring chemical technology to a country foreign to the supplier of the technology. The extent of the disaster would have been lessened, however, if Union Carbide had designed the plant with

smaller (though more) tanks to store MIC as it had been required to do in France.

The government of India required the Bhopal plant to be operated entirely by Indian workers. Hence, Union Carbide at first took admirable care in training plant personnel, flying them to its West Virginia plant for intensive training. It also had teams of U.S. engineers make regular on-site safety inspections. But in 1982, financial pressures led Union Carbide to relinquish its supervision of safety at the plant, but it retained general financial and technical control. The last inspection by a team of U.S. engineers occurred that year, two years before the explosion, despite the fact that the team had warned of many of the hazards that contributed to the disaster.

During the years after 1982, safety practices eroded. One source of the erosion was personnel: high turnover of employees, failure to properly train new employees, and low technical preparedness of the local labor pool. Workers handling pesticides, for example, learned more from personal experience than from study of safety manuals about the dangers of the pesticides. But even after suffer-

[5] Paul Shrivastava, *Bhopal, Anatomy of a Crisis* (Cambridge, MA: Ballinger, 1987). See also Gary Stix, "Bhopal: A Tragedy in Waiting," *IEEE Spectrum*, June 1989, pp. 47–50; Larry Everest, *Behind the Poison Cloud: Union Carbide's Bhopal Massacre* (Chicago: Banner, 1985).

ing chest pains, vomiting, and other symptoms, they would sometimes fail to wear safety gloves and masks because of high temperatures caused by lack of air-conditioning in the plant.

The other source of eroding safety practices was the move away from U.S. standards (contrary to Carbide's written policies) toward lower Indian standards. By December of 1984, several extreme hazards, in addition to many smaller ones, were present. (See Figure 9.1.)

First, the tanks storing the MIC gas were overloaded. Carbide's manuals specified they were never to be filled to more than 60 percent of capacity; in emergencies, the extra space could be used to dilute the gas. The tank that caused the problem was in fact more than 75 percent full.

Second, a standby tank that was supposed to be kept empty for use as an emergency dump tank already contained a large amount of the chemical.

Third, the tanks were supposed to be refrigerated to make the chemical less reactive if trouble should arise. But the refrigeration unit had been shut down five months before the accident as a cost-cutting measure, making tank temperatures three to four times what they should have been.

According to one account, a disgruntled employee unscrewed a pressure gauge on a storage tank and inserted a hose into it. He knew and intended that the water he poured into the tank would do damage, but he did not know it would cause such immense damage. According to another account, a relatively new worker had been instructed by a new supervisor to flush out some pipes and filters connected to the chemical storage tanks. Apparently the worker properly closed valves to isolate the tanks from the pipes and filters being washed, but he failed to insert the required safety disks to back up the valves in case they leaked. (He knew that valves leaked, but he did not check for leaks: "It was not my job." The safety disks were the responsibility of the maintenance department, and the position of second-shift supervisor had been eliminated.) Lawyers and their experts whose tasks include pinpointing blame, and engineers who design similar plants, need to know exactly what caused the pressure in the tank to rise, but for the purposes of our present discussion it is sufficient to be concerned about the subsequent failure of the plant's safety systems. By the time the workers noticed a gauge showing the mounting pressure and began to feel the sting of leaking gas, they found their main emergency procedures unavailable. The primary defense against gas leaks was a vent-gas scrubber designed to neutralize the gas. It was shut down (and was turned on too late to help), because it was assumed to be unnecessary during times when production was suspended.

The second line of defense was a flare tower that would burn off escaping gas missed by the scrubber. It was inoperable because a section of the pipe connecting it to the tank was being repaired. Finally, workers tried to minimize damage by spraying water 100 feet into the air. The gas, however, was escaping from a stack 120 feet high.

Within two hours, most of the chemicals in the tank had escaped to form a deadly cloud over hundreds of thousands of people in Bhopal. As was common in India, desperately poor migrant laborers had become squatters—by the tens of thousands—in the vacant areas surrounding the plant. They had come with hopes of finding any form of employment, as well as to take advantage of whatever water and electricity was available.

Virtually none of the squatters had been officially informed by Union Carbide or the Indian government of the danger posed by the chemicals being produced next door to them. (The only voice of caution was that of a concerned journalist, Rajukman Keswani, who had written articles on the dangers of the plant and had posted warnings: "Poison Gas. Thousands of Workers and Millions of Citizens are in Danger.") There had been no emergency drills, and there were no evacuation plans: The scope of the disaster was greatly increased because of total unpreparedness.

"When in Rome"

What, in general, are the moral responsibilities of multinational corporations, like Union Carbide and General Motors, and their engineers? One tempting view is that corporations and employees are merely obligated to obey the laws and dominant customs of the host country: "When in Rome do as the Romans do." This view is a version of ethical relativism, the claim that actions are morally right within a particular society when (and only because) they are approved by law, custom, or other conventions of that society. Ethical relativism, however, is false because it implies moral absurdities. For example, it would justify horrendously low safety standards, if that were all a country required. Laws and conventions are not morally self-certifying. Instead, they are always open to criticism in light of moral reasons concerning human rights, the public good, duties to respect people, and virtues.

An opposite view would have corporations and engineers retain precisely the same practices endorsed at home, never making any adjustments to a new culture. This view is a version of ethical absolutism, the claim that moral principles have no justified exceptions and that what is morally true in one situation is true everywhere else.

Absolutism is false because it fails to take account of how moral principles can come into conflict, forcing some justified exceptions. Absolutism also fails to take account of the many variable facts.

These considerations led us in Chapter 2 to endorse ethical relationalism: Moral judgments are and should be made in relation to factors that vary from situation to situation, usually making it impossible to formulate rules that are both simple and absolute. Moral judgments are contextual in that they are made in relation to a wide variety of factors—including the customs of other cultures. Note that relationalism only says that foreign customs are morally relevant. It does not say they are automatically decisive or self-authoritative in determining what should be done. This crucial difference sets it apart from ethical relativism.

Relationalism, we should add, is also consistent with ethical pluralism, the view that there is more than one justifiable moral perspective. In particular, there may be a number of morally permissible variations in formulating, interpreting, and applying basic moral principles. Not all rational and morally concerned people must see all specific moral issues the same way. This is as true in thinking about multinational corporations as it is in more everyday issues where we recognize that reasonable people can see moral issues differently and still be reasonable.

International Rights

If moral values are open to alternative interpretations, are there nevertheless some minimal standards that must be met? Let us respond to this question within the framework of rights ethics. A human right, by definition, is a moral entitlement that places obligations on other people to treat one with dignity and respect. If it makes sense at all, it makes sense across cultures, thereby providing a standard of minimally decent conduct that corporations and engineers are morally required to meet.

How can this general doctrine of human rights be applied practically, to help us understand the responsibilities of corporations doing business in other countries? In a pioneering book, The Ethics of International Business, Thomas Donaldson formulates a list of "international rights," human rights that are implied by, but more specific than, the most abstract human rights to liberty and fairness. These international rights have great importance and are often put at risk. Their exact requirements must be understood contextually, depending on the particular traditions and economic resources available in particular societies. Just as "ought implies can," rights do not require the impossible, and they also apply only within structured so-

cieties that provide a framework for understanding how to fairly distribute the burdens associated with them.

Donaldson suggests there are 10 such international rights.[6]

1. The right to freedom of physical movement.
2. The right to ownership of property.
3. The right to freedom from torture.
4. The right to a fair trial.
5. The right to nondiscriminatory treatment (freedom from discrimination on the basis of such characteristics as race or sex).
6. The right to physical security.
7. The right to freedom of speech and association.
8. The right to minimal education.
9. The right to political participation.
10. The right to subsistence.

These are human rights; as such, they place restrictions on how multinational corporations may act in other societies, even when those societies do not recognize the rights in their laws and customs. For example, the right to nondiscriminatory treatment would make it wrong for corporations to participate in discrimination against women and racial minorities even though this may be a dominant custom in the host country. Again, the right to physical security requires supplying protective goggles to workers running metal lathes, even when this is not required by the laws of the host country.

Although these rights have many straightforward cross-cultural applications, they nevertheless need to be applied contextually to take into account some features of the economy, laws, and customs of host countries. Not surprisingly, many difficulties and gray areas arise. One type of problem concerns the level of stringency required in matters such as degrees of physical safety at the workplace. Workers in less economically developed countries are often willing to take greater risks than would be acceptable to workers in the United States. Here Donaldson recommends applying a "rational empathy test" to determine if it is morally permissible for a corporation to participate in the practices of the host country: Would citizens of the home country find the practice acceptable if their home country were in circumstances economically similar to those of the host country? For example, in determining whether a certain degree of pollution is acceptable for a

[6] Thomas Donaldson, *The Ethics of International Business* (New York: Oxford University Press, 1989), p. 81.

U.S. company with a manufacturing plant located in India, the U.S. company would have to decide whether the pollution level would be acceptable under circumstances where the United States had a comparable level of economic development.

A second, quite different, type of problem arises where the practice is not so directly linked to economic factors, as in racial discrimination. Here Donaldson insists that unless one can do business in the country without engaging in practices that violate human rights, then corporations must simply leave and go to other countries.

Discussion Topics

1. Following the disaster at Bhopal, Union Carbide argued that officials at its U.S. corporate headquarters had no knowledge of the violations of Carbide's official safety procedures and standards. This has been challenged as documents were uncovered showing they knew enough to have warranted inquiry on their part, but let us assume they were genuinely ignorant. Would ignorance free them of responsibility for all aspects of the disaster?

2. Export of hazardous technologies, such as the manufacture of asbestos, to less-developed countries is motivated in part by cheaper labor costs, but another factor is that workers are willing to take greater risks. How does Donaldson's view apply to this issue?

   Also, do you agree with Richard De George's view that taking advantage of this willingness need not be unjust exploitation if several conditions are met: (1) Workers are informed of the risks. (2) They are paid more for taking the risks. (3) The company takes some steps to lower the risks, even if not to the level acceptable for U.S. workers.[7]

   How would you assess Union Carbide's handling of worker safety? Take into account the remarks of an Indian worker interviewed after the disaster. The worker was then able to stand only a few hours each day because of permanent damage to his lungs. During that time, he begged in the streets while he awaited his share of the legal compensation from Union Carbide. When asked what he would do if offered work again in the plant knowing what he knew now, he replied: "If it opened again tomorrow I'd be happy to take any

job they offered me. I wouldn't hesitate for a minute. I want to work in a factory, any factory. Before 'the gas' [disaster] the Union Carbide plant was the best place in all Bhopal to work."[8]

3. During 1972 and 1973, the president of Lockheed, A. Carl Kotchian, authorized secret payments totalling around $12 million beyond a contract to representatives of Japan's Prime Minister Tanaka. Later revelations of the bribes helped lead to the resignation of Tanaka and also to new laws in this country forbidding such payments. In 1995, long after Tanaka's death, the agonizingly slow trial and appeals process came to an end as Japan's Supreme Court reaffirmed the guilty verdicts, but so far no one has been jailed and the case appears to have had little recent impact on business and politics in Japan.

   Mr. Kotchian believed at that time it was the only way to assure sales of Lockheed's TriStar airplanes in a much-needed market. In explaining his actions, Mr. Kotchian cited the following facts.[9] (1) There was no doubt in his mind that the only way to make the sales was to make the payments. (2) No U.S. law at the time forbade the payments. (3) The payments were financially worthwhile, for they totaled only 3 percent of an expected $430 million income for Lockheed. (4) The sales would prevent Lockheed layoffs, provide new jobs, and thereby benefit workers' families and their communities as well as the stockholders. (5) He himself did not personally initiate any of the payments, which were all requested by Japanese negotiators. (6) In order to give the TriStar a chance to prove itself in Japan, he felt he had to "follow the functioning system" of Japan. That is, he viewed the secret payments as the accepted practice in Japan's government circles for this type of sale.

   (a) Drawing on the distinctions made in this section and Chapter 2, explain the several senses in which someone might claim that how Mr. Kotchian ought to have acted is a "relative" matter. Which of these senses, in your view, would yield true claims and which false?

   (b) Present and defend your view about whether Mr. Kotchian's actions were morally justified. In doing so, apply utilitarianism, rights ethics, and other ethical theories that you see as relevant.

[7] Richard T. De George, "Ethical Dilemmas for Multinational Enterprise: A Philosophical Overview," in *Ethics and the Multinational Enterprise,* ed. W. Michael Hoffman, Ann E. Lange, and David A. Fedo (New York: University Press of America, 1986).

[8] Fergus M. Bordewich, "The Lessons of Bhopal," *Atlantic Monthly,* March 1987, pp. 30–33.
[9] Carl Kotchian, "The Payoff: Lockheed's 70-day Mission to Tokyo," *Saturday Review,* July 9, 1977.

4. In 1977, the Foreign Corrupt Practices Act was signed into law, largely based on the Lockheed scandal. It makes it a crime for American corporations to accept payments from, or to offer payments to, foreign governments for the purpose of obtaining or retaining business, although it does not forbid "grease" payments to low-level employees of foreign governments (such as clerks) that are part of routine business dealings. Critics are urging repeal of the act because there is no question that it has adversely affected U.S. corporations trying to compete with countries that do not forbid paying business extortion. Is damage to profits a sufficient justification for repealing the act?

5. Since Nigeria became a member of OPEC (the Organization of Oil-Producing Countries) in 1970, the country's oil boom has led to increased corruption, lower living standards for the poor, and much political instability. Foreign oil companies (among whom Shell Oil has received much notoriety) are accused of having disregarded the safety and livelihood of local people when drilling and laying pipelines. The Ogony people in particular have protested, but in vain. Acquaint yourself with the happenings (then and now), and describe what you feel should be the role of foreign oil companies in a country such as Nigeria.

6. The World Trade Organization (WTO) was established to oversee trade agreements, enforce trade rules, and settle disputes. Some troublesome issues have arisen when WTO has denied countries the right to impose environmental restrictions on imports from other countries. Thus, for example, the United States may not impose a ban on fish caught with nets that can endanger other sealife such as turtles or dolphins, while European countries and Japan will not be able to ban imports of beef from U.S. herds injected with antibiotics. Investigate the current disputes and discuss how such problems may be resolved, not overlooking the fact that now a multinational company covering countries A and B has an opportunity to pressure A to relax environmental regulations under the guise of reduced export opportunities to country B, and vice versa regarding exports from B to A. Among other sources, you may wish to consult a contribution by Ralph Nader and Lori Wallach to a book on globalization.[10]

9.2 | ENVIRONMENTAL ETHICS

The Commons and a Livable Environment

Aristotle noted long ago that we tend to be thoughtless about things we do not own individually and that seem to be in unlimited supply. William Foster Lloyd was also an astute observer of this phenomenon. In 1833, he described what the ecologist Garrett Hardin would later call "the tragedy of the commons."[11]

Lloyd observed that cattle in the common pasture of a village were more stunted than those kept on private land. The common fields were themselves more worn than private pastures. His explanation began with the premise that individual farmers are understandably motivated by self-interest to enlarge their herd by one or two cows, especially given that each act taken by itself does negligible damage. Yet the combined effects of all the farmers behaving this way is the (tragic) overgrazing of the pasture so as to damage the good of everyone.

The same kind of competitive, unmalicious, but unthinking exploitation arises with all natural resources held in common: air, land, forests, lakes, oceans, endangered species, and, indeed, the entire biosphere. Hence, the tragedy of the commons remains a powerful image in thinking about environmental challenges in today's era of increasing population and decreasing natural resources. Its very simplicity, however, belies the complexity of many environmental issues. Here we cite several illustrations of that complexity, before turning to various kinds of solutions and philosophical perspectives on the environment.

Cases

Acid Rain—A Surprising Turn of Events   Normal rain has a pH of 5.6, but the typical rain in the northeastern areas of North America is now 3.9 to 4.3. This is 10 to 100 times more acidic than it should be, about as acidic as lemon juice. In addition, the snow-melt each spring releases huge amounts of acid that were in frozen storage during the winter months. Soil that contains natural buffering agents counteracts the acids. But large parts of the northeastern United States and eastern Canada lack sufficient natural buffers to counteract additional onslaughts.

The results? "Acid shock" from snow-melt is thought to have caused annual mass killings of fish. Longer-term effects of the acid harm fish eggs and food sources. Deadly quantities of aluminum, zinc, and many other metals

---

[10] Ralph Nader and Lori Wallach, "GATT, NAFTA, and the Subversion of the Democratic Process," ch. 8 in *The Case Against the Global Economy* (San Francisco: Sierra Club Books, 1996).

[11] Garrett Hardin, *Exploring New Ethics for Survival* (New York: Viking, 1968), p. 254.

leached from the soil by the acid rain also take a toll as they wash into streams and lakes. It was observed that in the higher elevations of the Adirondack Mountains, more than half the lakes that were once pristine can no longer support fish. Hundreds of other lakes were dying in the United States and Canada. Forests were also dying, and larger animals were suffering dramatic decreases in population, while some farmlands and drinking-water sources were being damaged.

These results occurred during only a few decades. It is believed that North America was just slightly behind Scandinavia, where thousands of lakes have been "killed" by acid rain. In both locations, the cause is now clear: the burning of fossil fuels that release large amounts of sulfur dioxide (SO_2)—the primary culprit—and nitrogen oxides (NOX). In both instances major sources of the pollutants are located hundreds and even thousands of miles away, with winds supplying a deadly transportation system to the damaged ecosystems. Much of Sweden's problem, for example, is traceable to the industrial plants of England and northern Europe. Acid rain problems in Canada and the northeastern United States derive in large measure from the utilities of the Ohio Valley, the largest source of sulfur dioxide pollution in this country. As we know now, pollution does not stop at national borders, necessitating international control of it.

Much remains to be learned about the mechanisms involved in the processes pictured in Figure 9.2. It is still im-

possible to link specific sources with specific damage. More research into shifting wind patterns and the air transport of acids is needed. Nor is there a reliable estimate of current damage. For example, many believe that microorganisms in soil are being affected in ways that are potentially devastating, but no one knows for sure. Groundwater is undoubtedly being polluted, but it is unclear what that means for human health. Much underground water currently being used was deposited by rainfall over a hundred years ago, and current acid rain may have its main effects on underground water a century from now. Effects on human food sources are also largely unknown. In some areas, certain trees do well; perhaps for them the acid rain acts as a fertilizer.

The good news is that acid rain is now being battled effectively in the United States. When Congress was debating amendments to the Clean Air Act in 1990, industry claimed costs of $3 to $7 billion a year to meet the first stage of reductions by the year 2000, and another $7 to $25 billion a year with the more stringent limits thereafter. But then Congress agreed to let operators of coal burning plants decide on their own how to cut sulphur dioxide emissions: by selecting scrubbers of their own choice or by using costlier cleaner-burning coal, and by trading reductions beyond those required. Thus, plants not performing so well could buy emissions credit from the cleaner plants, with dollars changing hands when these plants were not owned by the same company.

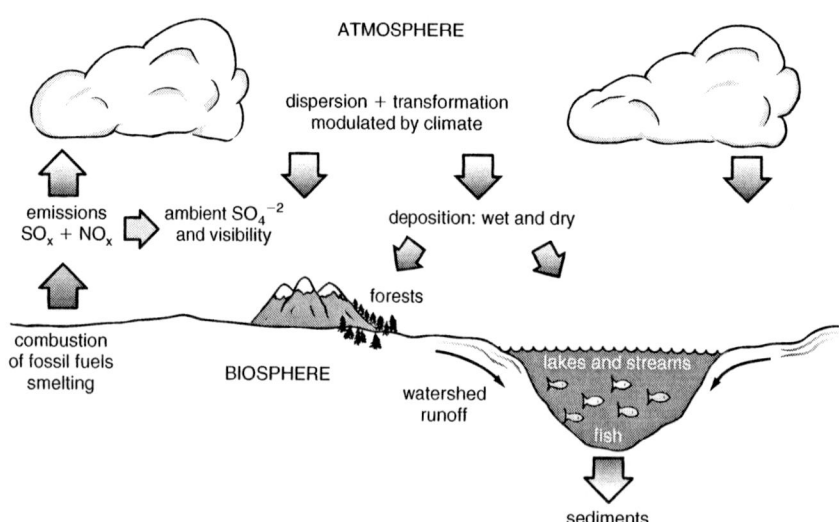

Figure 9.2 Acid deposition: sources and affected ecosystems
(National Research Council, *Acid Deposition*, Long-Term Effects [Washington, DC: National Academy of Sciences, 1986] p. 11. Diagram used with permission of National Academy Press. Also see Donald D. Adams and Walter P. Page, eds., *Acid Deposition* [New York: Plenum, 1985].)

All along, the Environmental Protection Agency (EPA) had estimated much lower costs for higher emission reductions, very much to industry's chagrin, but the final results exceeded even EPA's best hopes: by 1996, after the first two years of Phase I limits, emissions dropped to 35 percent below the legal limit at a cost of $0.8 billion a year! And EPA's cost estimate for Phase II (years 2000–2010) has dropped to an annual cost of $2.5 billion. It is hoped that this approach, also tried in variations by some European countries, may be useful in also combating greenhouse emissions worldwide.

Other Atmospheric Effects

Other examples can be given of amorphous patterns of ecological damage, like those of acid rain. Worldwide use of fossil fuels by industrial nations is causing a buildup of carbon dioxide in the atmosphere, which could result in a greenhouse effect damaging the entire earth. Some foreseeable effects are changing weather patterns and melting of polar ice-caps leading to raised water lines along ocean coasts. Already observable is an increased bacterial growth rate, as reported by the Physicians for Social Responsibility. Similarly, damage to the protective ozone layer of the earth's atmosphere resulting from the release of chlorofluorocarbons (CFCs) is related to technological products used by the populations of those same nations. (Where not now prohibited, CFCs such as Freon are used as refrigerants and as propellants in aerosol spray cans.) And rivers amass pollutants as they wind their way through several states or countries, eventually to dump their toxic contents into an ocean. The Rhine is such a river, and the North Sea, now a "special protection area," is such an ocean.

PCBs and Kanemi's Rice Oil

This case of a tainted supply of cooking oil is instructive because even with patients and health officials fully cooperating, a mysterious illness was difficult to define and the offending ingredient extremely hard to trace. How much more effort would it require then to conduct a similar investigation on animals in the wild or in the oceans! Yet their existence has been severely threatened in locations where the same kind of hardy ingredients, polychlorinated biphenyls or PCBs, have been carelessly discharged for decades as industrial waste. A search of web-sites will reveal that this is a world-wide problem and that humans can be affected by eating contaminated fish.

PCBs first appeared as by-products in oil refining processes around the turn of the century and were later recognized to be suitable for use as heat-transfer fluids and as insulating oils in transformers and capacitors. The feathers of stuffed birds in museums indicate that PCB had escaped into the environment before 1914, and from the 1930s to 1950s, producers of PCB oil (e.g., Monsanto) and manufacturers using it (e.g., Westinghouse and GE) were noticing skin rashes and other ailments afflicting workers who handled the oil. The extent of its spread was not recognized until the 1960s when researchers from the coast of Sweden to the coast of California showed that PCB had found its way through the food chain into fish and seabirds and was endangering their natural, biological reproduction. Then, in 1968, following an incident in Japan, the world's public health establishment began to examine PCBs in earnest. Let us pick up the PCB related tragedy in Japan after recalling briefly the sad state of environmental protection there in the 1960s.

A decade of rapid industrial growth in Japan had taken its toll on the environment. City dwellers fell ill from air pollution. Some rivers were covered with dead fish floating on the surface. The dreadful Minamata disease, from mercury pollution of a nearby bay by the Chisso Company, was continuing to fell its victims (there would be 46 deaths and 75 disabilities by 1970). In May, the itai-itai ("ouch-ouch") disease, which is painful and causes bone crumbling, was the first illness to be designated by the Japanese government as pollution-generated. Observed on and off since 1920 at various sites, the latest outbreak was placed at the doors of a Mitsui mining and smelting facility that let cadmium escape into a river, from which the

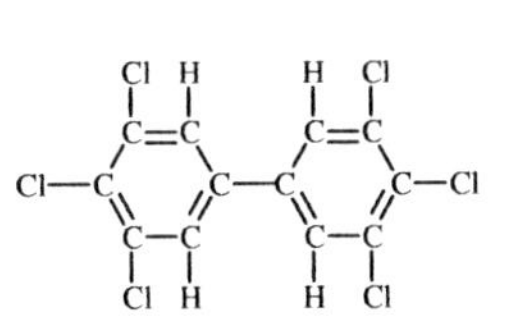

3,3',4,4'-Hexachlorobiphenyl

Figure 9.3 A typical PCB $C_{12}H_4C_{12}$

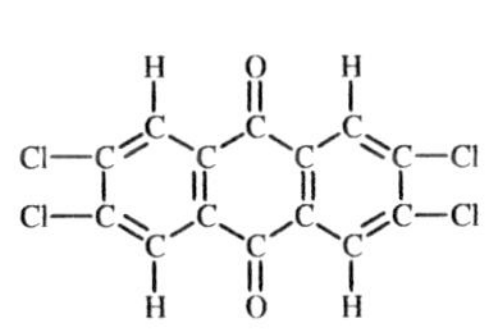

$C_{12}H_4O_2C_{14}$

2,3,7,8-tetrachlorodibenzo-p-dioxin

Figure 9.4 A typical dioxin, PCB's nasty cousin

pollutant found its way into the food chain via rice paddies and rice. The thalidomide drug tragedy was still on everyone's mind. The severely malformed children of women who had taken that drug during their pregnancies were still not cared for by either the Dainippon Pharmaceutical Company or the government, and schools refused to accept them.[12]

Then, in the summer of 1968, a disease of unknown origin appeared in southern Japan. Victims suffered from disfiguring skin acne and discoloration, fatigue, numbness, respiratory distress, vomiting, and loss of hair. Eventually 10,000 people were stricken and some died. What was the cause? An investigation of 121 cases was conducted, with 121 healthy individuals matched to the victims by age and sex being used as a control group. All 242 were questioned regarding their diets, personal habits, and places of work. When it was discovered that the only significant difference between the two groups was in the amount of fried foods eaten, the disease was traced to rice oil produced by the Kanemi Company.

It took another seven months to find the specific agent in the oil. Autopsies performed on victims revealed the presence of PCBs. Oil made from rice bran at the Kanemi plant was heated at low pressure to remove objectionable odors. The heating pipes were filled with hot Kanechlor, a PCB-containing fluid, but the pipes were corroding and tiny pinholes in them allowed PCBs to leak directly into the oil. In fact, Kanemi had been in the habit of replenishing about 27 kg of lost Kanechlor a month for some time.

There are other, less direct paths by which the extremely toxic PCBs can reach humans. For example, the Kanemi rice oil had also been used as an additive to chicken feed. In early 1968, one million chickens were given this feed and half the chickens died. In the United States, 140,000 chickens were slaughtered in the state of New York on one occasion when data collected by the Campbell's Soup Company revealed more than the permitted level of PCBs in chickens raised by certain growers. The source of PCBs was found to be plastic bakery wrappers mixed in with ground-up stale bread from bakeries used as feed. On another occasion PCBs leaked from a heating system into fishmeal in a North Carolina pasteurization plant. About 12,000 tons of fishmeal were contaminated and 88,000 chickens, already fed this fishmeal, had to be destroyed when the product was recalled.

PCBs were not used only in heat exchangers and electrical equipment. They were also good as hydraulic fluid and plastics additives. But they are no longer considered suitable for such applications, or any other where they can find their way into the environment. Their hardiness accounts for the fact that they were found in the oceans in larger quantities than DDT, although they entered the oceans initially in much lesser amounts. Even a total shutdown of all possible sources of PCB contamination would not result in a rapid reduction in its presence in the environment. Meanwhile production is continuing worldwide. It stood at 57 million pounds a year before 1989. Now that the U.S. has banned PCB applications, the production is still 22 million pounds a year. Information on PCBs is collected by the advocacy group COPA (Coalition Opposed to PCB Ash in Monroe County, Indiana) on its website www.org/library.

Too Little Water for the Everglades   The great marshes of southern Florida have attracted farmers and real estate developers since the beginning of the century. When drained, they present valuable ground. From 1909 to 1912, a fraudulent land development scheme was attempted in collusion with the U.S. Secretary of Agriculture. Arthur Morgan blew the whistle on that situation, jeopardizing not only his own position as a supervising drainage engineer with the U.S. Department of Agriculture, but also that of the head of the Office of Drainage Investigation. An attempt to drain the Everglades was made again by a Florida governor from 1926 to 1929. Once more, Arthur Morgan, this time in private practice, stepped in to reveal the inadequacy of the plans and thus discourage bond sales.

But schemes affecting the Everglades did not end then. Beginning in 1949, the U.S. Army Corps of Engineers started diverting excess water from the giant Lake Okeechobee to the Gulf of Mexico to reduce the danger of flooding to nearby sugar plantations. As a result, the Everglades, lacking water during the dry season, were drying up. A priceless wildlife refuge was falling prey to humanity's appetite. In addition, the diversion of waters to the Gulf and the ocean also affected human habitations in southern Florida. Cities that once thought they had unlimited supplies of fresh groundwater found they were pumping salt water instead as ocean waters seeped in.

Southern Florida is a complex environmental unit with a delicate balance. Any intrusion by human engineering must be seen as an experiment that must be conducted with great care. Unfortunately, too many public agencies view any change in plans as unacceptable once a course has been charted. As Arthur Morgan points out in his book *Dams and Other Disasters*, the Corps was particularly

---

[12] Jun Ui, ed., *Polluted Japan* (Tokyo: Jishu-Koza Citizens' Movement, 1972); Nobuko Iijima, *Pollution Japan* (New York: Pergamon, 1979).

prone to such an attitude, which was fostered by the crisis-oriented training at West Point Military Academy,[13] akin to "group-think." Crises of natural origin (such as floods) may, of course, still befall the Everglades, but then they will gradually recover as they have done by themselves for ages, provided human incursions have not been too severe.

These cases barely hint at the many environmental issues that arise in engineering practice, but they suffice to set a backdrop for distinguishing some ways of addressing environmental and international trade issues. Here we take note of four of these ways, all of which are essential dimensions of workable solutions: industry leadership, governments, market mechanisms, and individuals' commitments.

Corporations: Environmental Leadership

These cases barely hint at the multitude of alarming developments that lead many to speak of an environmental crisis. The good news, however, is that a wide consensus now exists about the importance of environmental issues and the need for concerted action by industry, government, revised market mechanisms, and individuals.

In the present climate, it is simply good business for a corporation to be perceived by the public as environmentally responsible, indeed as a leader in finding creative solutions. Compaq Computer Corporation is only one of a great many encouraging examples.[14] After being founded in 1982, it grew with astonishing success, making the Fortune 500 after only four years. As it did so, it made environmental commitments central to its mission, as recognized with a series of awards, including the 1997 World Environment Center Gold Medal for International Corporate Environmental Achievement.

Three features of Compaq's commitments are especially noteworthy as aspects of its "global" perspective on how its products affected the environment. First, Compaq developed a "life-cycle strategy" for its products that it dubbed "Design for Environment." Priorities were set for efficient use of resources, design of energy-efficient products, easy disassembly for recycling, and waste minimization. For example, it set a timetable for eliminating CFC emissions in its manufacturing process that was ahead of government requirements, and then met its goal two years ahead of schedule.

Second, Compaq developed unified standards that would apply throughout its operations. This was no minor feat, given that Compaq not only markets its products in over 100 countries, but also has subsidiaries in dozens of countries in North America, Latin America, Europe, the Middle East, Africa, and Asia. Rather than exploiting lower standards in other countries as an excuse to engage in cost savings, Compaq established consistent policies that serve as an exemplar for other companies and industries.

Third, in choosing its suppliers, Compaq places a high priority on companies with a record of environmental concern. Doing so tends to serve its business interests, since some of its costs are shifted to suppliers who already factor in part of the life-cycle concerns. But it also expresses Compaq's genuine and systematic commitments to make environmental responsibility a priority.

Fortunately Compaq is not alone in these efforts. IBM, for instance, also has an extensive (and extensively advertised) computer recycling program. An example of a government run system is Norway's effort to collect computers before they are discarded so they can be refurbished and donated to schools, or they are recycled.

Government: Enabled Natural Disasters and Technology Assessment

Nature's onslaughts can threaten communities and their infrastructures, but the mere occurrence of a hurricane, a tsunami or flood, an earthquake or a landslide, a volcanic eruption or a brush fire, does not in itself constitute a "disaster" or an inevitable "act of God." A disaster is the result of not being prepared for unusual yet predictable natural events. There are four sets of measures communities can take to avert or mitigate disasters without necessarily "stopping" nature.

One set of defensive measures consists of restrictions or requirements imposed on human habitat. For instance, homes should not be built in flood plains, homes in prairie country should have tornado shelters, hillsides should be stabilized to prevent landslides, structures should be able to withstand earthquakes and heavy weather, roof coverings should be made from nonflammable materials, and roof overhangs should be fashioned so flying embers will not be trapped. These are not nightmarish regulations, but merely reminders to developers and builders to do what their profession expects them to do anyway.

Another set of measures consists of strengthening—better yet, duplicating or arranging into grids—the life lines for essential utilities such as water (especially for fire fighting) and electricity. The third category encompasses special purpose defensive structures that would include dams, dikes, breakwaters, avalanche barriers, and

[13] Arthur E. Morgan, *Dams and Other Disasters* (Boston: Porter Sargent, 1971), pp. 370–89.
[14] Noel M. Tichy, Andrew R. McGill, and Lynda St. Clair, eds., *Corporate Global Citizenship* (San Francisco: New Lexington Press, 1997), pp. 230–44.

means to keep flood waters from damaging low lying sewage plants placed where gravity will take a community's effluents.

A fourth and final set of measures should assure "safe exit" in the form of roads designed as escape routes, structures designated as emergency shelters, adequate clinical facilities, and agreements with neighboring communities for sharing resources in emergencies.

Enabled Natural Disasters

The measures we have cited as examples will not avert emergencies, but they can prevent such emergencies from turning into disasters. Not to take precautions is tantamount to acting as an "enabler" of disaster, much like a person who lacks the foresight or the will to keep alcoholic beverages from a drunkard spouse or partner.

Unfortunately the lessons that could be learned from earlier disasters are often shrugged aside by a disbelief that it could occur again—"Lightning never strikes twice in the same place," and "Another 100-year flood is about that far away,"—or by a belief that government would once more hand out disaster relief payments. For instance, the 1989 Loma Prieta earthquake revealed that the columns of elevated highways needed strengthening, but this has not been fully implemented ten years later. The weaknesses in steel structures found after the 1994 Northridge earthquake have in many cases still not been reported to building departments because the owners claim that any inspection reports they had commissioned should be the owners' private property and concern (and had better not become known to a potential buyer).

Meanwhile the Kobe (Hanshin) earthquake of January 1995 has shown that the infrequency of earthquakes in a particular region is no cause for complacency. In fact, local, regional, and national government agencies were poorly prepared to handle large disasters of any kind in that part of Japan. Even the burning of structural debris after the earthquake was marred by dioxin emanations from the mass of twisted PVC tubing. One outstanding feat, however, was the then almost completed Akashi Kaikyo bridge near Kobe, at 3910 m (with a 1991 m span) the world's longest suspension bridge. The earthquake moved the massive southern anchor and pier by 1 m, so only a few girders had to be modified. The total length is 3911 m now!

On August 16th, 1999 and earthquake hit Izmit and neighboring cities in northwestern Turkey, including Istanbul. The damage to structures and the resulting death toll in the tens of thousands were unusually heavy. And why? Because during a building boom, multistory apart-

ment houses had been built and inspected without serious attention to seismic dangers even though this region was known to be part of an active earthquake belt.

Technology Assessment

What precautionary measures are best? The answers are not easy to find as becomes evident when one considers the many well-intentioned but mishandled large projects of the past, ranging from locating dams or rechanneling rivers to the safe disposal of "spent" nuclear fuel.

Government laws and regulations are understandably the lightning rod in environmental controversies. Few would question the need for the force of law in setting firm guidelines regarding the degradation of the "commons." Yet, how much law, of what sort, and to what ends are matters of continual disagreement.

Until 1995, the U.S. Congress had an Office of Technology Assessment. It prepared studies on the social and environmental effects of technology in areas such as cashless trading (via bank card), nuclear war, health care, and pollution. At the federal and state levels, many large projects must be examined in terms of their environmental impact before they are approved. The purpose of all this activity is praiseworthy. But how effective can it be?

Engineers, it is often said, tend to find the right answers to the wrong questions. The economist Robert Theobald made the following comment on education:

> The university is ideally designed to insure that you remain certain that you know the answers to questions that other people posed long ago. The problem today is that the questions we should be answering are not yet known. Unfortunately the process required for discovering the right questions is totally different from the process of discovering the right answers.[15]

It should be quite apparent that it is not easy to know what questions to ask. And technology assessment and other forecasting methods suffer because of this.

When scientists conduct experiments, they endeavor to distill some key concepts out of their myriad observations. As shown in Figure 9.5, a funnel can be used to portray this activity. At the narrow end of the funnel, we have the current wisdom, the state of the art. Engineers use it to design and build their projects. These develop in many possible directions, as shown by the shape of the lower, inverted funnel. The difficult task of technology assessment and environmental impact analyses is to explore the extent

---

[15] Quoted in Charles A. Thrall and Jerold M. Starr, eds., *Technology, Power, and Social Change* (Lexington, MA: Lexington, 1972), p. 17.

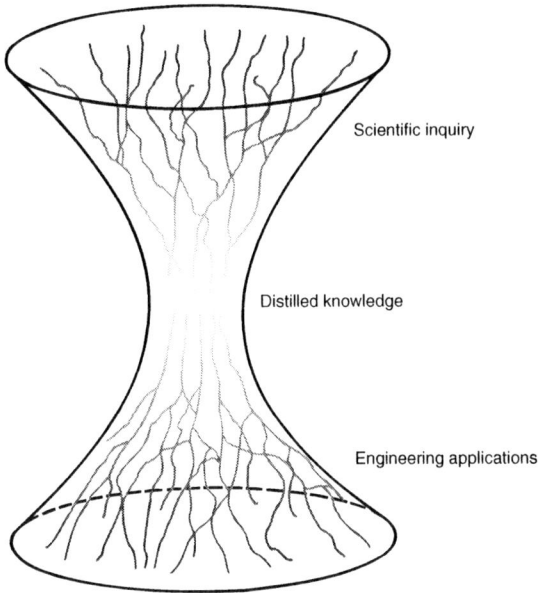

Figure 9.5 Distilling and applying knowledge.

of this spread and to separate the more significant among the possibly adverse effects.

The danger in any assessment of technology is that some serious risks can easily be overlooked while the studies and subsequent reports, properly authenticated by the aura of scientific methodology, assure the decision maker that nothing is amiss—or perhaps that perceived risks are more serious than they really are. We do not wish to belittle such efforts and we think that they are worthwhile, if only because of those questions they raise—and answers they uncover—that otherwise might not have surfaced. But there is a danger in believing that no further action is required once the reports have been approved and filed. Our contention remains that engineering must be understood as social experimentation and that the experiment continues, indeed enters a new phase, when the engineering project is implemented. Only by careful monitoring, which was well carried out by the Office of Technology Assessment, will it be possible to gather a more complete picture of the tangled web of effects encompassed in Figure 9.5 within the inverted, lower funnel.

Market Mechanisms: Internalizing Costs of Environmental Degradation

Democratic controls can take many forms beyond passing laws. One such option is internalizing costs of harm to the environment. When we are told how efficient and cheap many of our products and processes are—from agriculture to the manufacture of plastics—the figures usually include only the direct costs of labor, raw materials, and use of facilities. If we are quoted a dollar figure, it is at best an approximation of the price. The true cost would have to include numerous indirect factors such as the effects of pollution, the depletion of energy and raw materials, and social costs. If these, or an approximation of them, were internalized (added to the price), then those for whose benefit the environmental degradation had occurred could be charged directly for corrective actions.

As taxpayers are beginning to revolt against higher general levies, the method of having the user of a service or product pay for all its costs is gaining more favor. The engineer must join with the economist, the scientist, the lawyer, and the politician in an effort to find acceptable mechanisms for pricing and releasing products so that the environment is protected through truly self-correcting procedures rather than adequate-appearing yet often circumventable laws.

A working example is the tax imposed by governments in Europe on products and packaging that impose a burden on public garbage disposal or recycling facilities. The manufacturer prepays the tax and certifies so on the product or wrapper.

Fortunately, good design practices may in themselves provide the answers for environmental protection without added real cost. For example, consider the case of a lathe that was redesigned to be vibration-free and manufactured to close tolerances. It not only met occupational safety and health standards for noise, which its predecessor had not, but it also was more reliable and more efficient and had a longer useful life, thus offsetting the additional costs of manufacturing it.[16]

On a more ambitious level, one hopes that the many attempts to produce acceptable and affordable electric and other nonpolluting automobiles which industry is working on will soon succeed.

Individuals: Personal Commitments and Environmental Ethics

Individual engineers can make a difference. Although their actions are limited in a great many ways, they are uniquely placed to act as agents of change, as responsible

---

[16] Seymour Melman, "A Note on: Safety Improvements as a Zero Defect Problem," in *Designing for Safety: Engineering Ethics in Organizational Contexts*, ed. Albert Flores, (Troy, NY: Rensselaer Polytechnic Institute, 1982), p. 176.

experimenters. Doing so requires personal commitments based on a combination of moral concern and understanding of wider perspectives. Hence, we conclude this section with an overview of some of the environmental ethics that are currently being explored by concerned individuals in engineering and other professions.[17]

Human-Centered Environmental Ethics   We have been discussing "environmental ethics" as the study of the moral issues concerning the environment. Now we shift to "environmental ethics" in the sense of a general moral perspective on our moral responsibilities concerning the environment. Each of the ethical theories we have examined provides a foundation for environmental ethics. As human-centered environmental ethics, each focuses on the benefits of the natural environment to humans and the threats to human beings presented by the destruction of nature.

Thus, utilitarians emphasize that human pleasures and interests are linked to nature. Obviously, many of those pleasures and interests are linked to engineered products made from natural resources. In addition, however, we have aesthetic interests, as in the beauty of plants, waterfalls, and mountain ranges; recreational interests, as in hiking and backpacking in wilderness areas; scientific interests, especially in the study of "natural labs" of ecological preserves; and survival interests, which we have learned are linked directly to preserving the natural environment.

Duty ethics urges that respect for human life implies far greater concern for nature than has been traditionally recognized. Kant believed that we owe duties only to rational beings, which in his view excluded all nonhuman animals, although of course he did not have access to recent scientific studies showing striking parallels between humans and other primates. Nevertheless, he condemned callousness and cruelty toward conscious animals because he saw the danger that such attitudes would carry over to indecent treatment of persons. In any case, a duty-centered ethics would emphasize the need for conserving the environment because doing so is implied by respect for human beings who depend on it for their very existence.

Rights ethics urges that the basic right to life entails a right to a livable environment.[18] The right to a livable environment did not generally enter into people's thinking until the end of the twentieth century, at the time when pollution and resource depletion reached alarming proportions. Nevertheless, it is directly implied by the rights to life and liberty, given that these basic rights cannot be exercised without a supportive natural environment.

Finally, virtue ethics draws attention to the virtues of prudence, humility, appreciation of beauty, and gratitude toward the natural world that makes life possible,[19] and also the virtue of stewardship over resources that are needed for further generations.

Nature-Centered Ethics   All these human-centered ethics permit and indeed require a long-term view of conserving the environment. Not everything of importance within a human-centered ethics fits neatly into cost–benefit analyses with limited time horizons; much must be accounted for by means of constraints or limits that cannot necessarily be assigned dollar signs. Yet, some have argued that all versions of human-centered ethics are flawed and that we should widen the circle of things that have inherent worth, that is, value in themselves, independent of human desires and appraisals. Especially since 1979 when the new journal Environmental Ethics was founded, philosophers have reported on their explorations of a wide range of nature-centered ethics that, for example, affirm the inherent worth of all conscious animals, of all living organisms, or of ecosystems. Religious groups of all faiths have also given voice to their beliefs in human unity with nature.

Sentient-Centered Ethics   One version of nature-centered ethics recognizes all sentient animals as having inherent worth. Sentient animals are those that feel pain and pleasure and have desires. Thus, some utilitarians extend their theory (that right action maximizes goodness for all affected) to sentient animals as well as humans. Most notably, Peter Singer developed a utilitarian perspective in his influential book, Animal Liberation.[20] Singer insists that moral judgments must take into account the effects of our actions on sentient

[17] For helpful sources, see P. Aarne Vesilind and Alastair S. Gunn, *Engineering, Ethics, and the Environment* (New York: Cambridge University Press, 1998); Alastair S. Gunn and P. Aarne Vesilind, eds., *Environmental Ethics for Engineers* (Chelsea, MI: Lewis Publishers, 1986); Joseph R. DesJardins, *Environmental Ethics* 2nd ed. (Belmont, CA: Wadsworth, 1997); Susan J. Armstrong and Richard G. Botzler, eds., *Environmental Ethics* (New York: McGraw-Hill, 1993); Louis P. Pojman, *Environmental Ethics* (Boston: Jones & Bartlett, 1994); Donald VanDeVeer and Christine Pierce, eds., *The Environmental Ethics and Policy Book* (Belmont, CA: Wadsworth, 1994).

[18] William T. Blackstone, "Ethics and Ecology," in *Philosophy and the Environmental Crisis*, ed. William T. Blackstone (Athens, GA: University of Georgia Press, 1974).

[19] Thomas E. Hill Jr., "Ideals of Human Excellence and Preserving Natural Environments," *Environmental Ethics* 5 (1983).

[20] Peter Singer, *Animal Liberation*, rev. ed. (New York: Avon Books, 1990).

animals. Failure to do so is a form of discrimination akin to racism and sexism. He labels it "speciesism." Thus, in building a dam that will cause flooding to grasslands, engineers should take into account the impact on animals living there. Singer allows that sometimes animals' interests have to give way to human interests, but their interests should always be considered and weighed.

Singer does not ascribe rights to animals, and hence it is somewhat ironic that Animal Liberation has been called the bible of the animal rights movement. Other philosophers, however, do ascribe rights to animals. Most notably, Tom Regan contends that conscious creatures have inherent worth not only because they can feel pleasure and pain, but because, more generally, they are subjects of experiences who form beliefs, memories, intentions, and preferences and can act purposefully.[21] In his view, their status as subjects of experiments makes them sufficiently like humans to give them rights.

Singer and Regan tend to think of inherent worth as all-or-nothing. Hence, they think of conscious animals as deserving equal consideration. That does not mean they must be treated in the identical way we treat humans, but only that their interests should be weighed equally with human interests in making decisions. Other sentient-centered ethicists disagree. They regard conscious animals as having inherent worth, though not equal to that of humans.[22]

Biocentric Ethics   A life-centered ethics regards all living organisms as having inherent worth. Albert Schweitzer (1875–1965) set forth a pioneering version of this perspective under the name of "reverence for life."[23] He argued that the most fundamental feature of us is our will to live, by which he meant both a will to survive and a will to develop according to our innate tendencies. All organisms share these instinctive tendencies to survive and develop, and hence consistency requires that we affirm the inherent worth of all life.

Schweitzer often spoke of reverence for life as the fundamental excellence of character, and hence his view is a version of nature-centered virtue ethics. He refused to rank forms of life according to degrees of inherent worth,

but he believed that a sincere effort to live by the ideal and virtue of reverence for life would enable us to make inevitable decisions about when life must be maintained or has to be sacrificed. More recent defenders of biocentric ethics, however, have developed complex sets of rules for guiding decisions.[24]

Ecocentric Ethics   A frequent criticism of sentient-centered and biocentered ethics is that they are too individualistic, since they locate inherent worth in individual organisms. By contrast, ecocentered ethics locates inherent value in ecological systems. This approach was voiced by the naturalist Aldo Leopold (1887–1948), who urged that we have an obligation to promote the health of ecosystems: "A thing is right when it tends to preserve the integrity, stability, and beauty of the biotic community. It is wrong when it tends otherwise."[25] This "land ethic," as he called it, implied a direct moral imperative to preserve (leave unchanged), not just conserve (use prudently), the environment.

More recent defenders of ecocentric ethics have included within this holistic perspective an appreciation of human relationships. Thus, J. Baird Callicott writes that an ecocentric ethic does not "replace or cancel previous socially generated human-oriented duties—to family and family members, to neighbors and neighborhood, to all human beings and humanity."[26] That is, locating inherent worth in wider ecological systems does not cancel out or make less important what we owe to human beings. This way of thinking is in tune with Native American tribes who live in harmony with nature, based on a sense of respect and reverence for nature, while maintaining a primary respect for one another.

We have set forth these environmental ethics in connection with the reflections of individuals. Clearly, engineering would shut down if it had to grapple with theoretical disputes about human- and nature-centered ethics. Fortunately, at the level of practical issues the ethical theories often converge in the general direction for action, if not in all specifics. Just as humanity is part of nature, human-centered and nature-centered ethics overlap extensively in many of their practical implications.[27]

[21] Tom Regan, *The Case for Animal Rights* (Berkeley, CA: University of California Press, 1983).

[22] Mary Midgley, *Animals and Why They Matter* (Athens, GA: University of Georgia Press, 1984).

[23] Albert Schweitzer, "The Ethics of Reverence for Life," in *The Philosophy of Civilization,* trans. C. T. Campion (Buffalo, NY: Prometheus Books, 1987), pp. 307–29.

[24] Paul W. Taylor, *Respect for Nature* (Princeton, NJ: Princeton University Press, 1986).

[25] Aldo Leopold, *A Sand County Almanac* (New York: Ballantine, 1970), p. 262.

[26] J. Baird Callicott, "Environmental Ethics," in *Encyclopedia of Ethics,* vol. 1, ed. L. C. Becker (New York: Garland, 1992), pp. 313–14.

[27] James P. Sterba, "Reconciling Anthropocentric and Nonanthropocentric Environmental Ethics," in *Ethics in Practice,* ed. Hugh LaFollette (New York: Blackwell, 1997), pp. 644–56.

Discussion Topics

1. Exxon's 987-foot tanker Valdez was passing through Prince William Sound on March 24, 1989, carrying 50 million gallons of oil when it fetched up on Bligh Reef, tore its bottom, and spilled 11 million gallons of oil at the rate of a thousand gallons a second. This was one of the worst spills ever, not in quantity, but in its effect on a very fragile ecosystem. No human life was lost, but thousands of birds, sea otters, and other creatures died who were stuck in the oil or had fed on the poisoned carcasses.

    Discuss how each of the human-centered and nature-centered ethical theories would interpret the moral issues involved in this case.

2. Consider the following example of environmental side effects cited by Garrett Hardin:

    > *The Zambesi River . . . was dammed . . . to create the 1700-square-mile Lake Kariba. The effect desired: electricity. The "side-effects" produced: (1) destructive flooding of rich alluvial agricultural land above the dam; (2) uprooting of long-settled farmers from this land to be resettled on poorer hilly land that required farming practices with which they were not familiar; (3) impoverishment of these farmers . . . [and various other social disorders]; (6) creation of a new biotic zone along the lake shore that favored the multiplication of tsetse flies.[28]

    Similar problems have occurred when dams were built in the United States and when the Aswan Dam was erected on the Nile. One might ask if the original purpose may not itself begin to look like merely a side effect. If so, Hardin asks, can we never do anything? Describe under what conditions you think a dam such as the one on the Zambesi River should be built and operated. To whom is the engineer in charge of its construction ultimately responsible?

3. Write an essay on one of the following topics: Why Save Endangered Species?; Why Save the Everglades?; What are Corporations' Responsibilities Concerning the Environment? In your essay discuss the following question: What ethical theory would you apply to our relation to the environment?[29]

4. In part II of Faust, Goethe has Faust feverishly planning and supervising a huge land reclamation project. It is fueled by tricky concepts and devices, from paper money to steam engines—note that mephisto remains close at hand—but nature intervenes with tricks of its own. This is how eco-economist Hans Christoff Binswanger comments on Faust's predicament:

    > The real danger is that Faust—modern man—will not acknowledge the need for careful planning to forestall such damage as he pushes on relentlessly, not seeing what is going on around him. Goethe symbolizes this blind irresponsibility by Faust's loss of eyesight. . . . Hence mankind compounds its natural limitations—its inability to fully understand nature's complexity—with a blindness induced by hubris.[30]

    How do you interpret Goethe's use of the project to reclaim land from the sea?

---

[28] Garrett Hardin, *Exploring New Ethics for Survival* (New York: Viking, 1968), p. 68.

[29] See Nicholas Rescher, "Why Save Endangered Species?" in *Unpopular Essays and Technological Progress* (Pittsburgh: University of Pittsburgh Press, 1980), pp. 79–92; Bryan G. Norton, *Why Preserve Natural Variety?* (Princeton, NJ: Princeton University Press, 1987).

[30] Hans Christoff Binswanger, "The Challenge of Faust," SCIENCE, v. 281, 31 July 1998, pp. 640–641. See also Binswanger's book *Money and Magic: A Critique of the Modern Economy in Light of Goethe's Faust* (University of Chicago Press, 194).

Managing Project Teams

Chapter

Teamwork 10

Reflection

Think about a really effective team that you've been a member of, a team that accomplished extraordinary things and perhaps was even a great place to be. Start by thinking about teams in an academic, professional, or work setting. If no examples come to mind, then think about social or community-based teams. If again you don't conjure up an example, then think about sports teams. Finally, if you don't come up with a scenario from any of these contexts, then simply imagine yourself as a member of a really effective team. OK, got a picture of the team in mind? As you recall (or imagine) this highly effective team experience, try to extract the specific characteristics of the team. What was it about the team that made it so effective? Please make a list.

Look over the list you made in the above Reflection. Did you preface your list with "It depends"? The characteristics of an effective team depend, of course, on the purpose of the team. In large measure, they depend on the team's task goals (those concerning what the team is to do) and maintenance goals (those concerning how the team functions). Michael Schrage (1991) states emphatically:

> [P]eople should understand that real value in the sciences, the arts, commerce, and, indeed one's personal and professional lives, comes largely from the process of collaboration. What's more, the quality and quantity of meaningful collaboration often depend upon the tools used to create it . . . Collaboration is a *purposive* relationship. At the heart of collaboration is a desire or need to: solve a problem, create, or discover something. (p. 34) Within a set of constraints—expertise, time, money, competition, conventional wisdom. (p. 36)

Let's assume that it's a team that has both task and maintenance goals, since most effective teams not only have a job to do (a report to write, a project to complete, a presentation to give, etc.) but also a goal of getting better at working with one another.

I've used the Reflection above with hundreds of faculty and students in workshop and classroom settings. Here is a typical list of the characteristics of effective teams:

Good participation	Common goal
Respect	Sense of purpose
Careful listening	Good meeting facilitation
Leadership	Empowered members
Constructively managed conflict	Members take responsibility
Fun, liked to be there	Effective decision making

10.1 | DEFINITION OF A TEAM

Katzenbach and Smith (1993) studied teams that performed at a variety of levels and came up with four categories. *Pseudo teams* are those that perform below the level of the average member. *Potential teams* don't quite get going but struggle along at or slightly above the level of the average member. *Real teams* perform quite well, and *high-performing teams* perform at an extraordinary level. Katzenbach and Smith then looked for common characteristics of real teams and high-performing teams. All real teams could be defined as follows: a small number of people with complementary skills who are committed to a common purpose, performance goals, and an approach for which they hold themselves mutually accountable. High-performing teams met all the conditions of real teams and, in addition, had members who were deeply committed to one another's personal growth and success.

Reflection

Now think about the groups that are being used in your engineering classes. Think about your most successful or effective group project experience. What were the characteristics of the group? What were the conditions? Are they similar to your most effective groups?

10.2 | TYPES OF LEARNING TEAMS

There is nothing magical about teamwork in engineering classes. For example, while some types of learning teams increase the quality of classroom life and facilitate student learning, others hinder student learning and create disharmony and dissatisfaction with classroom life. To use team-

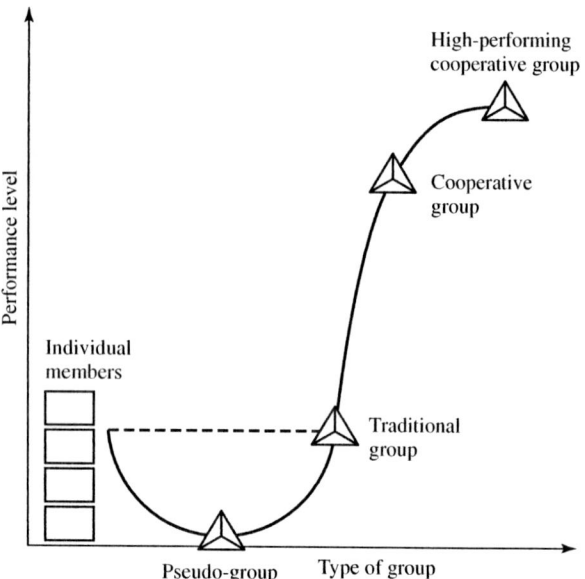

Figure 10.1 Group performance

work effectively, you must know what is and what is not a desirable characteristic.

When you choose to use (or are asked or required to use) instructional groups, you must ask yourself, "What type of group am I involved in?" Figure 10.1 and the following may be helpful in answering that question.

Pseudo Learning Group

Students in a pseudo learning group are assigned to work together but they have no interest in doing so. They believe they will be evaluated by being ranked from the highest performer to the lowest performer. While on the surface students talk to each other, under the surface they are competing. Because they see each other as rivals who must be defeated, they block or interfere with each other's learning, hide information from each other, attempt to mislead and confuse each other, and distrust each other. Students would achieve more if they were working alone.

Traditional Classroom Learning Group

Students in a traditional classroom learning group are assigned to work together and accept that they must do so. Assignments are structured, however, so that very little joint work is required. Students believe that they will be evaluated and rewarded as individuals, not as members of the group. They interact primarily to clarify how assignments are to be done. They seek each other's information, but have no motivation to teach what they know to their groupmates. Helping and sharing are minimized. Some students loaf, seeking a free ride on the efforts of their more conscientious groupmates. The conscientious members begin to feel exploited and therefore do less. The sum of the whole is more than the potential of some of the members, but the more hard-working and conscientious students would perform higher if they worked alone.

Cooperative Learning Group

Students in cooperative learning groups are assigned to work together and, given the complexity of the task and the necessity for diverse perspectives, they are relieved to do so. They know that their success depends on the efforts of all group members. The group format is clearly defined. First, the group goal of maximizing all members' learning provides a compelling common purpose that motivates members to roll up their sleeves and accomplish something beyond their individual achievements. Second, group members hold themselves and each other accountable for doing high-quality work to achieve their mutual goals. Third, group members work face-to-face to produce joint results. They do real work together. Students promote each other's success through helping, sharing, assisting, explaining, and encouraging. They provide both academic and personal support based on their commitment to and concern for each other. Fourth, group members are taught teamwork skills and are expected to use them to coordinate their efforts and achieve their goals. Both task and maintenance (team-building) skills are emphasized. All members share responsibility for providing leadership. Finally, groups analyze how effectively they are achieving their goals and how well members are working together. There is an emphasis on continual improvement of the quality of learning and teamwork processes. A recent guide to success in active learning is available in the book *Striving for Excellence in College* (Browne and Keeley, 1997).

High-Performance Cooperative Learning Group

A high-performance cooperative learning group meets all the criteria for being a cooperative learning group and outperforms all reasonable expectations, given its membership. What differentiates the high-performance group from the ordinary cooperative learning group is the level of commitment members have to each other and to the group's success. Jennifer Futernick, who is part of a high-performing, rapid-response team at McKinsey & Company, calls the emotional binding together of her team-

mates a form of love (Katzenbach and Smith, 1993). Ken Hoepner of the Burlington Northern Intermodal Transport Team (also described in Katzenbach and Smith) stated: "Not only did we trust each other, not only did we respect each other, but we gave a damn about the rest of the people on this team. If we saw somebody vulnerable, we were there to help." Members' mutual concern for each other's personal growth enables high-performance cooperative groups to perform far above expectations, and also to have lots of fun. The bad news about extraordinarily high-performance cooperative learning groups is that they are rare. Most groups never achieve this level of development.

10.3 | GROUPS AND TEAMS

I've been using the terms *group* and *team* interchangeably and I will continue to do so throughout this book. The traditional literature focuses on groups, while recently some writers have been making distinctions between groups and teams. Katzenbach and Smith (1993) summarize the major differences between working groups and teams (see Figure 10.2).

From your perspective, are there any surprises in Figure 10.2? Many students emphasize the importance of a strong leader, but Katzenbach and Smith indicate that real teams, as opposed to working groups, have shared leadership roles. Also notice that the literature on high-performance teams indicates that they are composed of members with complementary skills; that is, they're diverse.

10.4 | IMPORTANCE OF DIVERSITY

Often we must work with people who are different from us or difficult to work with but whose skills, talents, expertise, and experience are essential to the project. Working with a diverse group may seem impossible at times, but look at the example of Phil Jackson, former head coach of the Chicago Bulls basketball team. Can you imagine more a more diverse group than one made up of Dennis Rodman, Michael Jordan, and Scottie Pippin? Phil Jackson is an expert at managing diversity. Ethnic diversity is increasing in the workplace and in the broader society. Many predict that today's ethnic minorities will grow as a proportion of the population; in fact, Hispanics are predicted to become the majority in the near future.

Diversity has many faces, including preferred learning style (visual, auditory, kinesthetic); social background and experience; ethnic and cultural heritage; gen-

Working Group	Team
Strong, clearly focused leader	Shared leadership roles
Individual accountability	Individual and mutual accountability
Purpose the same as the broader organizational mission	Specific team purpose that the team itself delivers
Individual work-products	Collective work-products
Runs efficient meetings	Encourages open-ended discussion and active problem-solving meetings
Measures its effectiveness indirectly by its influence on others	Measures performance directly by assessing collective work-products
Discusses, decides, and delegates	Discusses, decides, and does real work together

Figure 10.2 Not all groups are teams: How to tell the difference.
Source: Katzenbach and Smith, 1993.

der; and sexual orientation. The evidence from effective groups is that diversity is important; that is, the better a group represents the broader community, the more likely it is to make significant, creative, and acceptable contributions. Participating in and managing diverse groups are not always easy tasks, since diverse groups usually encompass a wide range of ideas and priorities. The following are some steps you can take in learning to manage diverse groups more effectively (Cabanis, 1997; Cherbeneau, 1997):

1. Learn skills for working with all kinds of people.
2. Stress that effective teams are diverse.
3. Stress the importance of requirements.
4. Emphasize performance.
5. Develop perspective-taking skills, i.e. putting yourself in other's shoes.
6. Respect and appreciate alternative perspectives.

The Chicago Bulls' former head coach Phil Jackson has said, "Good teams become great ones when the members trust each other enough to surrender the 'me' for the 'we.'" His 1995 book (coauthored with Hugh Delehanty), *Sacred Hoops: Spiritual Lessons of a Hardwood Warrior,* offers terrific advice on organizing and managing extraordinarily high-performing teams.

10.4 | CHARACTERISTICS OF EFFECTIVE TEAMS

The research on highly effective teams both in the classroom (Johnson, Johnson, and Smith, 1991, 1998a, 1998b) and in the workplace (Bennis and Biederman, 1997; Hargrove, 1998; Katzenbach and Smith, 1993; Schrage, 1991, 1995) reveals a short list of the characteristics of effective teams:

1. *Positive interdependence.* The group focuses on a common goal or single product.
2. *Individual and group accountability.* Each person takes responsibility for both his or her work and the overall work of the group.
3. *Promotive interaction.* The members do real work, usually face-to-face.
4. *Teamwork skills.* Each member has and practices effective communication (especially careful listening), decision making, problem solving, conflict management, and leadership.
5. *Group processing.* The group periodically reflects on how well the group is working, celebrates the things that are going well, and problem-solves the things that aren't.

Teams have become commonplace in engineering practice and are making inroads in engineering education. The immense literature on teams and teamwork, ranges from very practical guides (e.g., Scholtes, Joiner, and Streibel, 1996; Brassand, 1995) to conceptual and theoretical treatises (e.g., Johnson and Johnson, 1991; Hackman, 1990). Check out one of these to broaden and deepen your understanding of teamwork. Four books were highlighted in this chapter—*Shared Minds: The New Technologies of Collaboration* (Schrage, 1991); *The Wisdom of Teams: Creating the High-Performance Organization* (Katzenbach and Smith, 1993); *Organizing Genius: The Secrets of Creative Collaboration* (Bennis and

Biederman, 1997); and *Mastering the Art of Creative Collaboration* (Hargrove, 1998). These four books focus on extraordinary teams—teams that perform at unusually high levels and whose members experience accomplishments through synergistic interaction that they rarely experience in other settings. They provide lots of examples and insights into high-performance teams. Katzenbach and Smith, for example, give the following advice for building team performance:

- Establish urgency and direction.
- Select members based on skill and potential, not personalities.
- Pay attention to first meeting and actions.
- Set clear rules of behavior.

- Set some immediate performance-oriented tasks and goals.
- Challenge the group regularly with fresh information.
- Spend *lots* of time together.
- Exploit the power of positive feedback, recognition, and reward.

Effective teamwork is not easy to accomplish. Engineering professor Douglas J. Wilde said, "It's the soft stuff that's hard, the hard stuff is easy." (Leifer, 1997) However if you work at it, and continue to study and learn about effective teamwork, you will very likely have many positive team experiences (and save yourself a lot of grief). Chapter 3 presents specific skills and strategies needed for effective teamwork.

Questions

1. What are the characteristics of effective teams? How do you help promote them?

2. Where and how have teamwork skills been taught or emphasized to you? In school? social groups? professional groups? your family? Describe two or three instances where teamwork skills were emphasized.

3. How is increasing ethnic diversity affecting project teams? What are some strategies for effectively participating on and managing diverse teams?

4. Students often remark, "But groups in school are different from groups in the workplace." The remark is delivered as a reason for not using groups in school. Is it a valid excuse? Summarize the major differences between groups in school and groups in the workplace. How are these differences beneficial or harmful to the work of the group? What are some things that you can do to improve the school groups?

Exercises

1. Check out a study of teams that have performed at extraordinary levels. Some of the books listed in the references for this chapter have terrific stories of stellar teams. You may want to check the library or do an electronic search of the literature. Summarize the features of extraordinary teams. How do they compare with the list provided in this chapter? Remember, this is a dynamic area of research with lots of new books and articles appearing each year.

2. Look for opportunities to participate on a superb team. Make a plan for participating on a high-performance team.

3. Study the diversity of teams in your school or workplace, and note strategies for recognizing, valuing, and celebrating diversity.

References

Bennis, Warren, and Patricia Biederman. 1997. *Organizing genius: The secrets of the creative organization.* Reading, MA: Addison-Wesley.

Brassand, Michael. 1995. *The team memory jogger: A pocket guide for team members.* Madison, WI: GOAL/QPC and Joiner Associates.

Browne, M. Neil, and Stuart Keeley. 1997. *Striving for excellence in college.* Upper Saddle River, NJ: Prentice-Hall.

Cabanis, Jeannette. 1997. Diversity: This means you. *PM Network* 11(10): 29–33.

Cherbeneau, Jeanne. 1997. Hearing every voice: How to maximize the value of diversity on project teams. *PM Network* 11(10): 34–36.

Hackman, J. R. 1990. *Groups that work (and those that don't): Creating conditions for effective teamwork.* San Francisco: Jossey-Bass.

Hargrove, Robert. 1998. *Mastering the art of creative collaboration.* New York: McGraw-Hill.

Jackson, Phil, and Hugh Delehanty. 1995. *Sacred hoops: Spiritual lessons of a hardwood warrior.* Hyperion.

Johnson, David W., and Frank P. Johnson. 1991. *Joining together: Group theory and group skills,* 4th ed. Englewood Cliffs, NJ: Prentice-Hall.

Johnson, David W., Roger T. Johnson, and Karl A. Smith. 1991. *Cooperative learning: Increasing college faculty instructional productivity,* Washington: ASHE-ERIC Reports on Higher Education.

———. 1998a. *Active learning: Cooperation in the college classroom,* 2nd ed. Edina, MN: Interaction Book Company.

———. 1998b. Maximizing instruction through cooperative learning. *ASEE Prism* 7(6): 24–29.

Katzenbach, Jon, and Douglas Smith. 1993. *The wisdom of teams: Creating the high-performance organization.* Cambridge, MA: Harvard Business School Press.

Leifer, L. 1997. Design team performance: Metrics and the impact of technology. In S. M. Brown & C. J. Seidner, eds., *Evaluating corporate training: Models and issues.* Kluwer Academic Publishers: 297–320.

Scholtes, Peter R., Brian L. Joiner, and Barbara J. Streibel. 1996. *The team handbook,* 2nd ed. Madison, WI: Joiner Associates.

Schrage, Michael. 1991. *Shared minds: The new technologies of collaboration.* New York: Random House.

———. 1995. *No more teams! Mastering the dynamics of creative collaboration.* New York: Doubleday.

Teamwork Skills and Problem Solving 11

Reflection

Have you been a member of a team that got the job done (wrote the report, finished the project, completed the laboratory assignment) but that ended up with the members hating one another so intensely they never wanted to see each other again? Most students have, and they find it very frustrating. Similarly, have you been a member of a team whose members really enjoyed one another's company and had a great time socially, but in the end hadn't finished the project? Again, most students have been a member of this type of group and they find it also a frustrating experience. Take a moment to recall your experiences with these two extremes of teamwork.

11.1 | IMPORTANCE OF TASK AND RELATIONSHIP

As noted in Chapter 10, to be most effective, groups need to do two things very well: accomplish the task and get better at working with one another. Both of these require leadership—not just from a single person acting as the leader but also from every member contributing to the leadership of the group. This chapter focuses on teamwork skills using a "distributed actions approach" to leadership. *Distributed actions* are specific behaviors that group members engage in to help the group accomplish its task or to improve working relationships. Napier and Gershenfeld (1973) summarize many of these behaviors (see Figure 11.1). Note the date—1973—which indicates that effective group work is not a new concept.

To realize the benefits of a team culture requires a change in management behavior, as shown in Figure 11.2. If the behaviors listed on the right-hand side of Figure 11.2 are not common in the groups you participate in, read on.

Group Task Roles	Group Maintenance Roles
Initiating	Encouraging
Seeking information	Expressing feelings
Giving information	Harmonizing
Seeking opinions	Compromising
Giving opinions	Facilitating communications
Clarifying	Setting standards or goals
Elaborating	Testing agreement
Summarizing	Following

Figure 11.1 Group task and maintenance roles
Source: Napier and Gershenfeld, 1973.

From	To
Directing	Guiding
Competing	Collaborating
Relying on rules	Focusing on the process
Using organizational hierarchy	Using a network
Consistency/sameness	Diversity/flexibility
Secrecy	Openness/sharing
Passive acceptance	Risk taking
Isolated decisions	Involvement of others
People costs	People assets
Results thinking	Process thinking

Figure 11.2 Management behavior change needed for team culture
Source: McNeill, Bellamy, and Foster, 1995.

11.2 | ORGANIZATION: GROUP NORMS

A common way to promote more constructive and productive teamwork is to have the teams create a set of guidelines for the group, sometimes called group norms. Take a minute and list some things (attitudes, behaviors, and so on) that you have found or think that would help a group be more effective. Then compare your list with the following two lists, both of which are from McNeill, Bellamy, and Foster (1995). The first was adapted from the Boeing Airplane Group's training manual for team members, and the second is from the Ford Motor Company.

Code of Cooperation

1. *Every* member is responsible for the team's progress and success.
2. Attend all team meetings and be on time.
3. Come prepared.
4. Carry out assignments on schedule.
5. Listen to and show respect for the contributions of other members; be an active listener.
6. *Constructively* criticize ideas, not persons.
7. Resolve conflicts constructively.
8. Pay attention; avoid disruptive behavior.
9. Avoid disruptive side conversations.
10. Only one person speaks at a time.
11. Everyone participates; no one dominates.
12. Be succinct; avoid long anecdotes and examples.
13. No rank in the room.
14. Respect those not present.
15. Ask questions when you do not understand.
16. Attend to your personal comfort needs at any time, but minimize team disruption.
17. Have fun.
18. ?

Ten Commandments: An Affective Code of Cooperation

- Help each other be right, not wrong.
- Look for ways to make new ideas work, not for reasons they won't.
- If in doubt, check it out. Don't make negative assumptions about each other.
- Help each other win, and take pride in each other's victories.
- Speak positively about each other and about your organization at every opportunity.
- Maintain a positive mental attitude no matter what the circumstances.
- Act with initiative and courage, as if it all depends on you.
- Do everything with enthusiasm; it's contagious.
- Whatever you want, give it away.
- Don't lose faith.
- Have fun.

Having an agreed-upon code of cooperation such as the ones listed above will help groups get started toward working effectively. However, if group members haven't developed the requisite communication, trust, loyalty, organization, leadership, decision-making procedures, and conflict management skills, then the group will very likely struggle or at least not perform up to its potential. One way a team can develop such a code is to create a *team charter*, which includes the following:

- Team name, membership, and roles.
- Team mission statement.
- Anticipated results (goals).
- Specific tactical objectives.
- Ground rules/guiding principles for team participation.
- Shared expectations/aspirations.

Team charters are typically created during a team meeting early in the project life cycle. Involvement of all team members in creating the charter helps build commitment of each to the project and other team members. A set of guidelines such as those listed above often help the team through this process.

11.3 | COMMUNICATION

Effective communication—listening, presenting, persuading—is at the heart of effective teamwork. The task and maintenance roles listed above all involve oral communication. Here are the listening skills emphasized in Arizona State University's Introduction to Engineering Design (McNeill, Bellamy & Foster, 1995):

> Stop talking.
>
> Engage in one conversation at a time.
>
> Empathize with the person speaking.
>
> Ask questions.
>
> Don't interrupt.
>
> Show interest.
>
> Concentrate on what is being said.
>
> Don't jump to conclusions.
>
> Control your anger.
>
> React to ideas, not to the speaker.
>
> Listen for what is not said; ask questions.
>
> Share the responsibility for communication.

Three listening techniques they recommend are:

Critical listening
- Separate fact from opinion.

Sympathetic listening.
- Don't talk—listen.
- Don't give advice—listen.
- Don't judge—listen.

Creative listening.
- Exercise an open mind.
- Supplement your ideas with another person's ideas and vice versa.

You may be wondering why so much emphasis on listening. The typical professional spends about half of his or her business hours listening and project managers may spend an even higher proportion of their time listening. Most people, however, are not 100 percent efficient in their listening. Typical listening efficiencies are only 25 percent (Taylor, 1998). The first list provides suggestions to help the listener truly hear what is being said and the second highlights that different situations call for different types of listening.

Reflection

Take a moment to think about listening skills and techniques. Do you listen in all three ways listed above? Which are you best at? Which do you need to work on?

11.4 | LEADERSHIP

A common notion is that leadership is a trait that some people are born with. Another common notion is that a person's leadership ability depends on the situation. There is an enormous literature on leadership, so I'll provide only insights that I've found useful. I'll also try to guide you to more reading and resources on the topic.

Individual and Group Reflection

What does it mean to lead a team? What does it take? Take a moment to reflect on the characteristics you admire most in a leader. Jot down 8 to 10 of them. Compare with your team.

Leadership authors Kouzes and Posner (1987, 1993) have asked thousands of people to list the characteristics of leaders they admire. Figure 11.3 lists the most common responses from their 1987 and 1993 studies. Many students and workshop participants express surprise at the listing of honesty as the characteristic mentioned most often. They say it's a given. Apparently honesty is not a given for many leaders in business and industry. In 1993, Kouzes and Posner also asked the respondents to list the most desirable characteristics of colleagues. Honest was number one again, with 82 percent selecting it. Cooperative, dependable, and competent were second, third, and fourth, with slightly over 70 percent of respondents selecting each.

Kouzes and Posner found that when leaders do their best, they challenge, inspire, enable, model, and encourage. They suggest five practices and 10 behavioral commitments of leadership.

Characteristic	1987 U.S. Percentage of People Selecting	1993 U.S. Percentage of People Selecting
Honest	83%	87%
Forward-looking	62	71
Inspiring	58	68
Competent	67	58
Fair-minded	40	49
Supportive	32	46
Broad-minded	37	41
Intelligent	43	38
Straightforward	34	34
Courageous	27	33
Dependable	32	32
Cooperative	25	30
Imaginative	34	28
Caring	26	27
Mature	23	14
Determined	20	13
Ambitious	21	10
Loyal	21	10
Self-controlled	13	5
Independent	13	5

Figure 11.3 Characteristics of admired leaders
Source: Kouzes and Posner, 1987, 1993.

Challenging the Process
1. Search for opportunities.
2. Experiment and take risks.

Inspiring a Shared Vision
3. Envision the future.
4. Enlist others.

Enabling Others to Act
5. Foster collaboration.
6. Strengthen others.

Modeling the Way
7. Set the example.
8. Plan small wins.

Encouraging the Heart
9. Recognize individual contributions.
10. Celebrate accomplishments.

Peter Scholtes, author of the best-selling book *The Team Handbook,* also wrote *The Leader's Handbook* (1998). He offers the following six "New Competencies" for leaders:

1. The ability to think in terms of systems and knowing how to lead systems.
2. The ability to understand the variability of work in planning and problem solving.
3. Understanding how we learn, develop, and improve; leading true learning and improvement.
4. Understanding people and why they behave as they do.
5. Understanding the interaction and interdependence between systems, variability, learning, and human behavior; knowing how each affects the others.
6. Giving vision, meaning, direction, and focus to the organization.

In addition to group norms, communication, and leadership, teamwork depends on effective decision making and constructive conflict management, described in the next two sections.

11.4 | DECISION MAKING

This section on decision making includes both strategies for decision making in groups and more general considerations for addressing ranking tasks.

Individual and Group Reflection

How do you typically make decisions in groups? Do you vote? Do you defer to the "expert"? Do you try to reach consensus? Take a moment to reflect on how the groups you participate in typically make decisions.

What did you come up with? Compare your reflection with those of other group members.

There are several approaches to making decisions in groups. Before exploring them, however, I suggest that you try a group decision-making exercise. Common exercises to assist in the development of teamwork skills, especially communication (sharing knowledge and expertise), leadership, and decision making are ranking tasks, such as the survival tasks, in which a group must decide which items are most important for survival in the desert, on the moon, or in some other difficult place. Ranking tasks are common in organizations that must select among alternative designs, hire personnel, or choose projects or proposals for funding.

My favorite ranking task for helping groups focus on communication, leadership, decision making, and conflict resolution is "They'll Never Take Us Alive." This exercise, which includes both individual and group decision making, is included at the end of this chapter. Do it now.

Group Reflection 1

How did your group make the decision? Did you average your individual rankings? Vote? Did you discuss your individual high and low rankings and then work from both ends toward the middle? Did you try to reach consensus? Were you convinced by group members who seem to have "expert" knowledge? Did you start with the number of fatalities for one of the activities and work from there?

Group refection 2

How well did your group work? What went well? What things could you do even better next time?

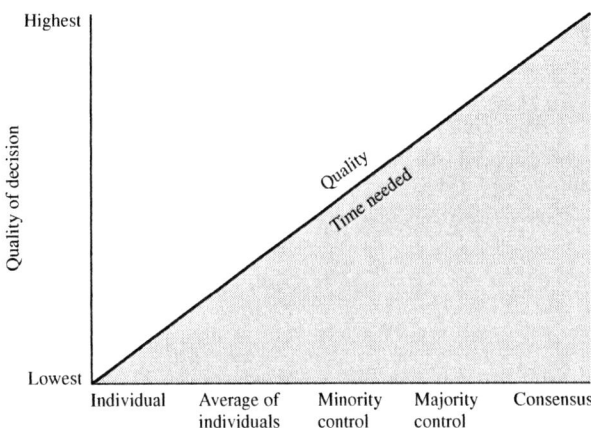

Figure 11.4 Decision type and quality

The method a group uses to make a decision depends on many factors, including how important the decision is, and how much time there is. Groups should have a good repertoire of decision-making strategies and a means of choosing the one that is most appropriate for the situation.

Several methods have been described in the literature for making decisions. One of my favorites is from David Johnson and Frank Johnson (1991). The authors list seven methods for making decisions:

1. *Decision by authority without discussion.* The leader makes all the decisions without consulting the group. It is efficient but does not build team member commitment to the decision.

2. *Expert member.* Group decision made by letting the most expert member decide for the group. The difficulty is often deciding who has the most expertise, especially when those with power or status in the group overestimate their expertise.

3. *Average of members' opinions.* Group decision based on average of individual group members' opinions.

4. *Decision by authority after discussion.* Group in which designated leader makes decision after discussion with the group. Effectiveness often depends on the listening skills of the leader.

5. *Minority control.* Two or more members who constitute less than 50 percent of the group often makes decisions by (a) acting as an executive committee or (b) special problem solving sub group.

6. *Majority control.* Decision by a majority vote is the most common method used in the U.S. Discussion occurs only until at least 51 percent of the members decide on a course of action.

7. *Consensus.* Consensus is probably the most effective method of group decision making, but it also

may take the most time. Perfect consensus is achieved when everyone agrees. A lesser degree of consensus is often accepted where everyone has had their say and will commit to the decision, but they may not completely agree with the decision.

They note that the quality of the decision and the time needed vary as a function of the number of people involved in the decision-making method, as shown in Figure 11.4.

David and Frank Johnson (1991) also list the following characteristics of effective decisions:

1. The resources of the group members are well used.

2. Time is well used.

3. The decision is correct, or of high quality.

4. The decision is put into effect fully by all the necessary members' commitments.

5. The problem-solving ability of the group is enhanced.

Group Reflection

How well did your group do on each of these five characteristics of effective decisions?

Typically, novice decision-making groups don't take full advantage of the skills and talents of their members, and they often struggle to get started. Some researchers report a series of stages in team development (e.g., forming, storming, norming, performing) and offer suggestions for working through each stage (Scholtes, Joiner, and Streibel, 1996). Also, if you ask a group to invest time and effort in making a decision it is very important that the decision be implemented (or very good rationale provided for why it wasn't implemented). There are few things more frustrating than to be asked to spend lots of time and effort on work that goes nowhere.

11.5 | CONFLICT MANAGEMENT

Conflict is a routine aspect of every project manager's job. *Conflict* is a situation in which an action of one person prevents, obstructs, or interferes with the actions of another person. On complex projects and tasks, highly talented and motivated people routinely disagree about the best ways to accomplish tasks and especially about how to deal with trade-offs among priorities. A conflict often is a moment of truth, since its resolution can follow either a constructive or a destructive path.

Individual Reflection

Write the word conflict in the center of a blank piece of paper and draw a circle around it. Quickly jot down all the words and phrases you associate with the word conflict by arranging them around your circle.

Review your list of associations and categorize them as positive, negative, or neutral. Count the total number of positive, negative, and neutral associations, and calculate the percentage that are positive. Did you have more than 90 percent positive?

Less than 5 percent of the people who have done this Reflection in my classes and workshops have had more than 90 percent positive associations. The majority, in fact, have had less than 50 percent positive associations. Many have had less than 10 percent positive.

The predominance of negative associations with conflict is one of the reasons conflict management is so difficult for project managers. Many people prefer to avoid conflict or to suppress it when it does arise. They become fearful, anxious, angry, or frustrated; consequently, the conflict takes a destructive path.

The goal of this section is to help you develop a set of skills and procedures for guiding conflict along a more constructive path. I'd like to begin by asking you to complete a questionnaire to assess how you typically act in conflict situation. The "How I Act in Conflict" question-

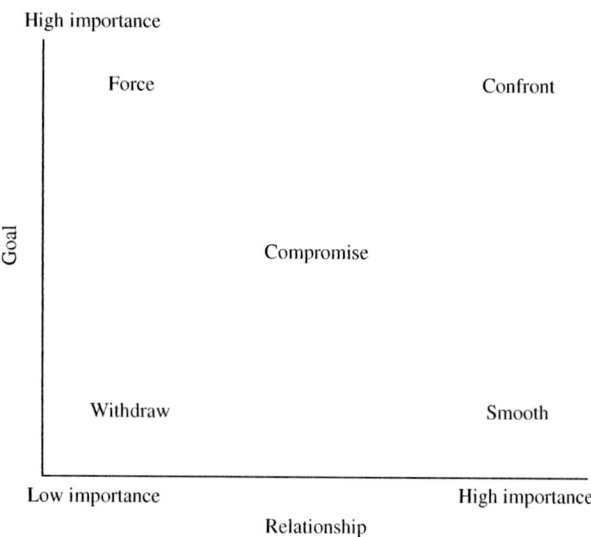

Figure 11.5   Blake and Mouton conflict model

naire is included as Exercise 2 at the end of this chapter. Take a few minutes to complete and score the questionnaire. Try to use professional conflicts and not personal conflicts as your point of reference.

Set the questionnaire aside for a few minutes and read Exercise 3, the Ralph Springer case study. Work through the exercise, completing the ranking form at the end.

Group Activity

Share and discuss each member's results from Exercise 2. Discuss each of the possible ways to resolve the conflict.

Then compare your individual responses from Exercise 2 to your rankings in Exercise 3. Note that each of the alternatives listed in Exercise 3 represents one of the five strategies on the scoring form in Exercise 2. Match the alternatives to the strategies they represent. Discuss similarities and differences in the order each group member would have used the strategies and the relative effectiveness of each.

The five conflict strategies shown in Exercise 2—withdrawal, forcing, smoothing, compromise, and confrontation—were formulated into a model for analyzing approaches to conflict by Blake and Mouton (1964). The authors used two axes to represent the conflict strategies: (1) the importance of the goal, and (2) the importance of the task. The placement of each of the five strategies according to this framework is shown in Figure 11.5.

The five strategies are described as follows:

1. *Withdrawal.* Neither the goal nor the relationship is important—you withdraw from the interaction.

2. *Forcing.* The task is important but not the relationship—use all your energy to get the task done.

3. *Smoothing.* The relationship is more important than the task. You want to be liked and accepted.

4. *Compromise.* Both task and relationship are important, but there is a lack of time—you *both* gain and lose something.

5. *Confrontation.* Task and relationship are equally important. You define the conflict as a problem-solving situation and resolve through negotiation.

Each of these strategies is appropriate under certain conditions. For example, if neither the goal nor the relationship is important to you, then often the best thing to do is withdraw. If the relationship is extremely important and the task is not so important (at the time), then smoothing is appropriate. In many conflict situations, both the task and the relationship are important. In these situations, confronting and negotiating often lead to the best outcomes.

A *confrontation* is the direct expression of one opponent's view of the conflict, and his or her feelings about it, and an invitation to the other opponent to do the same.

Guidelines for Confrontation

1. Do not "hit and run." Confront only when there is time to jointly define the conflict and schedule a negotiating session.

2. Openly communicate your feelings about and perceptions of the issues involved in the conflict, and try to do so in minimally threatening ways.

3. Accurately and fully comprehend the opponent's views and feelings about the conflict.

Negotiation is a conflict resolution process by which people who want to come to an agreement try to work out a settlement.

Steps in Negotiating a Conflict

1. Confront the opposition.
2. Define the conflict mutually.
3. Communicate feelings and positions.
4. Communicate cooperative intentions.
5. Take the other person's perspective.
6. Coordinate the motivation to negotiate.
7. Reach an agreement that is satisfactory to both sides.

Constructively resolving conflicts through a confrontation–negotiation process takes time and practice to perfect, but it's worth it. Conflicts that do not get resolved at

a personal level must be resolved at more time-consuming and costly levels—third-party mediation; arbitration; and, if all else fails, litigation.

Finally, here are some heuristics for dealing with conflicts in long-term personal and professional relationships:

1. Do not withdraw from or ignore the conflict.
2. Do not engage in "win–lose" negotiations.
3. Assess for smoothing.
4. Compromise when time is short.
5. Confront to begin problem-solving negotiations.
6. Use your sense of humor.

Remember that heuristics are reasonable and plausible, but not guaranteed. I suggest that you develop your own set of heuristics for dealing with conflict as well as for the other skills needed for effective teamwork. Some of my former students who now work as project managers emphasize during classroom visits that they spend a lot of time resolving conflicts—over meeting specifications, schedules, delivery dates, interpersonal differences among team members—and that most conflicts are dealt with informally.

11.6 | TEAMWORK CHALLENGES AND PROBLEMS

Reflection

What are some of the most common challenges and problems you've had working in groups? Please reflect for a moment. Make a list. Has a professor ever had you do this in your teams? If so, it's a clear indication that the professor understands the importance of group processing for identifying and solving problems.

What's on your list?

The challenges and problems you listed in the above reflection may have included the following:

- Members who don't show up for meetings or who don't show up prepared.
- Members who dominate the conversation.
- Members who don't participate in the conversation.
- Time wasted by off-task talk.
- Members who want to do the entire project because they don't trust others.
- Group meeting scheduling difficulties.
- No clear focus or goal.
- Lack of clear agenda, or hidden agendas.
- Subgroups excluding or ganging up on one or more members.

- Ineffective or inappropriate decisions and decision-making processes.
- Suppression of conflict or unpleasant flare-ups among group members.
- Members not doing their fair share of the work.
- Lack of commitment to the group's work by some members.

The problems listed above are commonly encountered by students (and professionals) working in groups. If they are not addressed they can turn a cooperative group into a pseudo group, as described in Chapter 2, where the group does worse than individuals working alone. If the challenges are addressed in a problem-solving manner, then the group is likely to perform at much higher levels (and the members will have a much more positive experience). The following process is widely used to address group problems.

Step 1: Identifying Challenges, Difficulties, and Barriers to Effective Group Work

- Reflect individually for a moment and start a list of challenges, barriers, or problems facing the group. Share the individual lists and create a joint list that includes at least one item from each group member.
- Do not solve (yet).
- Be realistic and specific.
- Work cooperatively.
- If more than one group is involved, list challenges, barriers, and problems for all groups on an overhead projector or flip chart.

Step 2: Addressing Barriers, Challenges, and Problems

1. Have each group or (if only one group is involved) each member select one item from the joint list.
2. Clarify: Make sure you have a common understanding of what the item means or represents.

3. Create three possible actions that will solve or eliminate the barrier.
4. Prioritize the possible solutions: Plan A, Plan B, Plan C.
5. Focus on what *will* work; be positive and constructive.
6. Implement the solutions; report back; celebrate and spread the ones that are effective.

Caveat: During implementation of group work expect some challenges, barriers, and problems. Doing so will help you recognize a roadblock when it appears. When it does appear, apply the appropriate parts of Step 2.

With one or more colleagues, develop three or more solutions. Implement one of these and then evaluate, replan, and retry.

The problem identification, problem formulation, and problem-solving format described above does not guarantee that your teamwork experiences will be free from troubles. But having a format for getting problems out on the table and then dealing with them in a problem-solving manner usually reduces the frustration and interference of group problems.

11.7 | REFLECTION: TEAMWORK

I've tried to address many of the highlights of effective teamwork and team problem solving, but I've barely scratched the surface. Hundreds of books and articles have been written on effective teamwork, and I've listed a few of my favorites in the reference section (see, e.g., Fisher, Rayner, and Belgard, 1995; Goldberg, 1995; Hackman, 1990; Katzenbach and Smith, 1993a, 1993b). As I mentioned earlier, one of the most widely used teamwork books is Scholtes, Joiner, and Streibel's (1996) *The Team Handbook*.

| Questions

1. What other skills besides those mentioned in this chapter do you feel are essential for successful groups? How about trust and loyalty, for example? I briefly dealt with trust and loyalty under the organization section, but you may want to emphasize them more. Check the references (e.g., David Johnson and Frank Johnson, 1991) for more. What other teamwork skills would you like to follow up on?

2. What are some of the strategies for developing a good set of working conditions in a group?
3. What are your reactions to the list of characteristics of admired leaders in Figure 11.3? Were you surprised by the high ranking of honesty?
4. Why is conflict central to effective teamwork and project work? What are some strategies for effectively managing conflict?

5. Keep a log of problems you've faced in working on project teams. How do the problems change over the life of the group?

6. The next time a problem occurs in a group, try the problem-solving process outlined in the chapter. How well did it work?

Exercises

1. They'll Never Take Us Alive

On the accompanying chart, in alphabetical order, are listed the top 15 causes of death in the United States in 1997. The data were taken from an annual review of death certificates. Your task is to rank the products and activities in order of the number of deaths they cause each year. Place the number 1 next to the one that causes the most deaths, the number 2 by the one that causes the second most deaths, and so forth. Then, write in your estimate of the number of fatalities each product or activity causes.

Group Tasks

1. After individuals have filled in the chart, determine one ranking for the group. (Do not worry yet about the estimates for the number of fatalities.)
2. Every group member must be able to explain the rationale for the group's ranking.

3. When your group finishes, and each member has signed the chart, (a) record your estimated number of fatalities in the U.S. for each, and then (b) compare your rankings and estimates with those of another group.

2. How I Act in Conflict

The proverbs listed in the accompanying table can be thought of as descriptions of some of the different strategies for resolving conflicts. Proverbs state conventional wisdom, and the ones listed here reflect traditional wisdom for resolving conflicts. Read each carefully. Using the scale provided, indicate how typical each proverb is of your actions in a conflict. Then score your responses on the chart at the end of the table. The higher the total score in each conflict strategy, the more frequently you tend to use that strategy. The lower the total score for each conflict strategy, the less frequently you tend to use that strategy.

Product or Activity	Ranking	Number of Fatalities
Accidents		
Alzheimer's disease		
Blood poisoning		
Cancer		
Diabetes		
Hardening of arteries		
Heart disease		
HIV and AIDS		
Homicide		
Kidney disease		
Liver disease		
Lung disease		
Pneumonia and influenza		
Stroke		
Suicide		

Sources: Office of the Surgeon General; National Center for Health Statistics.

5 = Very typical of the way I act in a conflict
4 = Frequently typical of the way I act in a conflict
3 = Sometimes typical of the way I act in a conflict
2 = Seldom typical of the way I act in a conflict
1 = Never typical of the way I act in a conflict

_____ 1. It is easier to refrain than to retreat from a quarrel.

_____ 2. If you cannot make a person think as you do, make him or her do as you think.

_____ 3. Soft words win hard hearts.

_____ 4. You scratch my back, I'll scratch yours.

_____ 5. Come now and let us reason together.

_____ 6. When two quarrel, the person who keeps silent first is the most praiseworthy.

_____ 7. Might overcomes right.

_____ 8. Smooth words make smooth ways.

_____ 9. Better half a loaf than no bread at all.

_____ 10. Truth lies in knowledge, not in majority opinion.

_____ 11. He who fights and runs away lives to fight another day.

_____ 12. He hath conquered well that hath made his enemies flee.

_____ 13. Kill your enemies with kindness.

_____ 14. A fair exchange brings no quarrel.

_____ 15. No person has the final answer, but every person has a piece to contribute.

_____ 16. Stay away from people who disagree with you.

_____ 17. Fields are won by those who believe in winning.

_____ 18. Kind words are worth much and cost little.

_____ 19. Tit for tat is fair play.

_____ 20. Only the person who is willing to give up his or her monopoly on truth can ever profit from the truths that others hold.

_____ 21. Avoid quarrelsome people, for they will only make your life miserable.

_____ 22. A person who will not flee will make others flee.

_____ 23. Soft words ensure harmony.

_____ 24. One gift for another makes good friends.

_____ 25. Bring your conflicts into the open and face them directly; only then will the best solution be discovered.

_____ 26. The best way of handling conflicts is to avoid them.

_____ 27. Put your foot down where you mean to stand.

_____ 28. Gentleness will triumph over anger.

_____ 29. Getting part of what you want is better than not getting anything at all.

_____ 30. Frankness, honesty, and trust will move mountains.

_____ 31. There is nothing so important that you have to fight for it.

_____ 32. There are two kinds of people in the world, the winners and the losers.

_____ 33. When someone hits you with a stone, hit him or her with a piece of cotton.

_____ 34. When both people give in halfway, a fair settlement is achieved.

_____ 35. By digging and digging, the truth is discovered.

3. Case Study—Ralph Springer

The following case gives you a chance to apply the Blake and Mouton (1964) conflict model to a hypothetical situation. Read the case carefully and then label each of the

Scoring

Withdrawal	Forcing	Smoothing	Compromise	Confrontation
1.	2.	3.	4.	5.
6.	7.	8.	9.	10.
11.	12.	13.	14.	15.
16.	17.	18.	19.	20.
21.	22.	23.	24.	25.
26.	27.	28.	29.	30.
31.	32.	33.	34.	35.
Total	Total	Total	Total	Total

Source: David Johnson and Roger Johnson, 1991.

possible actions from most to least effective and from most to least likely.

> You have been working as a project manager in a large company for some time. You are friends with most of the other project managers and, you think, respected by all of them. A couple of months earlier, Ralph Springer was hired as a supervisor. He is getting to know the other project managers and you. One of the project managers in the company, who is a friend of yours, confided in you that Ralph has been saying rather nasty things about your looks, the way you dress, and your personal character. For some reason you do not understand, Ralph has taken a dislike to you. He seems to be trying to get other project managers to dislike you also. From what you hear, there is nothing too nasty for him to say about you. You are worried that some people might be influenced by him and that some of your co–project managers are also beginning to talk about you behind your back. You are terribly upset and angry at Ralph. Since you have a good job record and are quite skilled in project management, it would be rather easy for you to get another job.

Rank each of the following five courses of action from 1 (most effective, most likely) to 5 (least effective, least likely). Use each number only once. Be realistic.

Effective Likely

_____ _____ I lay it on the line. I tell Ralph I am fed up with the gossip. I tell him that he'd better stop talking about me behind my back, because I won't stand for it. Whether he likes it or not, he is going to keep his mouth shut about me or else he'll regret it.

_____ _____ I try to bargain with him. I tell him that if he will stop gossiping about me I will help him get started and include him in the things other project managers and I do together. I tell him that others are angry about the gossiping and that it is in his best interest to stop. I try to persuade him to stop gossiping in return for something I can do.

_____ _____ I try to avoid Ralph. I am silent whenever we are together. I show a lack of interest whenever we speak, look over his shoulder and get away as soon as possible. I want nothing to do with him for now. I try to cool down and ignore the whole thing. I intend to avoid him completely if possible.

_____ _____ I call attention to the conflict between us. I describe how I see his actions and how it makes me feel. I try to begin a discussion in which we can look for a way for him to stop making me the target of his conversation and a way to deal with my anger. I try to see things from his viewpoint and seek a solution that will suit us both. I ask him how he feels about my giving him this feedback and what his point of view is.

_____ _____ I bite my tongue and keep my feelings to myself. I hope he will find out that the behavior is wrong without my saying anything. I try to be extra nice and show him that he's off base. I hide my anger. If I tried to tell him how I feel, it would only make things worse.

References

Blake, R. R., and J. S. Mouton. 1964. *The managerial grid.* Houston: Gulf Publishing Company.

Fisher, Kimball, Steven Rayner, and William Belgard. 1995. *Tips for teams: A ready reference for solving common team problems.* New York: McGraw-Hill.

Goldberg, David E. 1995. *Life skills and leadership for engineers.* New York: McGraw-Hill.

Hackman, J. R. 1990. *Groups that work (and those that don't): Creating conditions for effective teamwork.* San Francisco: Jossey-Bass.

Johnson, David W., and Frank P. Johnson, 1991. *Joining together: Group theory and group skills,* 4th ed. Englewood Cliffs, NJ: Prentice-Hall.

Johnson, David W., and Roger T. Johnson, 1991. *Teaching students to be peacemakers.* Edina, MN: Interaction Book Company.

Katzenbach, Jon R., and Douglas K. Smith. 1993a. *The wisdom of teams: Creating the high-performance organization.* Cambridge, MA: Harvard Business School Press.

————. 1993*b*. The discipline of teams. *Harvard Business Review* 71(2): 111–20.

Kouzes, J. M., and B. Z. Posner. 1987. *The leadership challenge: How to get extraordinary things done in organizations.* San Francisco: Jossey-Bass.

————. 1993. *Credibility: How leaders gain and lose it, why people demand it.* San Francisco: Jossey-Bass.

McNeill, Barry, Lynn Bellamy, and Sallie Foster. 1995. *Introduction to engineering design.* Tempe, Arizona: Arizona State University.

Napier, Rodney W., and Matti K. Gershenfeld. 1973. *Groups: Theory and experience.* Boston: Houghton Mifflin.

Scholtes, Peter R. 1998. *The leader's handbook: Making things happen, getting things done.* New York: McGraw-Hill.

Scholtes, Peter R., Brian L. Joiner, and Barbara J. Streibel. 1996. *The team handbook,* 2nd ed. Madison, WI: Joiner Associates.

Taylor, James. 1999. *A survival guide for project managers.* New York: AMACOM.

Chapter

Managing Projects

12

This chapter was developed in collaboration with Stephen Raab.

A manufacturer of microfilm imaging equipment approached the Eastman Kodak Company to design and supply the microfilm cartridges for use with a new machine under development (Figure 12.1). The target specifications were similar to previous products developed by the cartridge group at Kodak. However, in contrast to the usual 24-month development time, the customer needed prototype cartridges for demonstration at a trade show in just 8 months, and production was to begin 4 months later. Kodak accepted this challenge of cutting its normal development time in half and called its efforts the Cheetah project. Effective project management was crucial to the successful completion of the project.

For all but the simplest products, product development involves many people completing many different tasks. Successful product development projects result in high-quality, low-cost products while making efficient use of time, money, and other resources. *Project management* is

the activity of planning and coordinating resources and tasks to achieve these goals.

Project management activities occur during *project planning* and *project execution.* Project planning involves scheduling the project tasks and determining resource requirements. The project plan is first laid out during the concept development phase, although it is a dynamic entity and continues to evolve throughout the development process.

Project execution, sometimes called *project control,* involves coordinating and facilitating the myriad tasks required to complete the project in the face of inevitable unanticipated events and the arrival of new information. Execution is just as important as planning; many teams fail because they do not remain focused on their goals for the duration of the project.

This chapter contains five remaining sections. We first present the fundamentals of task dependencies and timing, along with three tools for representing relationships among project tasks. In the second section we show how these principles are used to develop an effective product development plan. In the third section we provide a set of guidelines for completing projects more quickly. After that, we discuss project execution, and finally we present a process for project evaluation and continuous improvement.

Figure 12.1 The Cheetah microfilm cartridge.
(Courtesy of Eastman Kodak Company.)

12.1 | UNDERSTANDING AND REPRESENTING TASKS

Product development projects involve the completion of hundreds or even thousands of tasks. This section discusses some of the fundamental characteristics of interacting tasks—the "basic physics" of projects. We also present three ways to represent the tasks in a project.

Sequential, Parallel, and Coupled Tasks

Figure 12.2 displays the tasks for three portions of the Cheetah project. The tasks are represented by boxes and

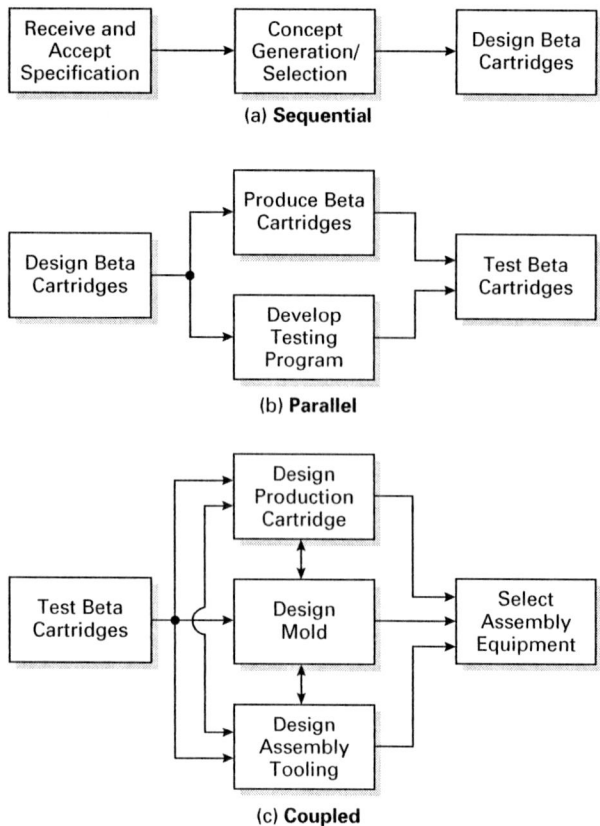

Figure 12.2 The three basic types of task dependencies: (a) sequential, (b) parallel, and (c) coupled.

other. The task on the right depends on the middle two tasks. We call the middle two tasks *parallel* because they are both dependent on the same task but are independent of each other. Figure 12.2(c) shows five development tasks, three of which are *coupled*. Coupled tasks are mutually dependent; each task requires the result of the other tasks in order to be completed. Coupled tasks either must be executed simultaneously with continual exchanges of information or must be carried out in an iterative fashion. When coupled tasks are completed iteratively, the tasks are performed sequentially with the understanding that the results are tentative and that each task will most likely be repeated one or more times until the team converges on a solution.

The Design Structure Matrix

A useful tool for representing and analyzing task dependencies is the *design structure matrix* (DSM). This representation was originally developed by Steward (1981) for the analysis of design descriptions and has more recently been used to analyze development projects modeled at the task level (Eppinger, 1994). Figure 12.3 shows a DSM for the 14 major tasks of the Cheetah project. (Kodak's actual plan included more than 100 tasks.)

In a DSM model, a project task is assigned to a row and a corresponding column. The rows and columns are named and ordered identically, although generally only the rows list the complete names of the tasks. Each task is defined by a row of the matrix. We represent a task's dependencies by placing marks in the columns to indicate the other tasks (columns) on which it depends. Reading across a row reveals all of the tasks whose output is required to perform the task corresponding to the row. Reading down a column reveals which tasks receive information from the task corresponding to the column. The

the information (data) dependencies among the tasks are represented by arrows. We refer to this representation as an *information-processing view* or a *data-driven perspective* of product development because most of the dependencies involve transfer of information (data) between the tasks. We say that task B is *dependent* on task A if an output of task A is required to complete task B. This dependency is denoted by an arrow from task A to task B.

Figure 12.2(a) shows three tasks, two of which are dependent on the output of another task. These tasks are *sequential* because the dependencies impose a sequential order in which the tasks must be completed. (Note that when we refer to tasks being "completed" sequentially, we do not necessarily mean that the later task cannot be started before the earlier one has been completed. Generally the later task can begin with partial information but cannot finish until the earlier task has been completed.) Figure 12.2(b) shows four development tasks. The middle two tasks depend only on the task on the left, but not on each

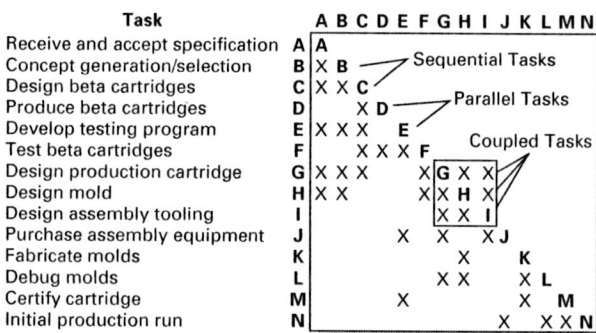

Figure 12.3 Simplified design structure matrix for the Kodak Cheetah project.

diagonal cells are usually filled in with dots or the task labels, simply to separate the upper and lower triangles of the matrix and to facilitate tracing dependencies.

The DSM is most useful when the tasks are listed in the order in which they are to be executed. In most cases, this order will correspond to the order imposed by sequential dependencies. Note that if only sequentially dependent tasks were contained in the DSM, then the tasks could be sequenced such that the matrix would be lower triangular; that is, no marks would appear above the diagonal. A mark appearing above the diagonal has special significance; it indicates that an earlier task is dependent on a later task. An above-diagonal mark could mean that two sequentially dependent tasks are ordered backward, in which case the order of the tasks can be changed to eliminate the above-diagonal mark. However, when there is no ordering of the tasks that will eliminate an above-diagonal mark, the mark reveals that two or more tasks are coupled.

Changing the order of tasks is called *sequencing* or *partitioning* the DSM. Simple algorithms are available for partitioning DSMs such that the tasks are ordered as much as possible according to the sequential dependencies of the tasks. Inspection of a partitioned DSM reveals which tasks are sequential, which are parallel, and which are coupled and will require simultaneous solution or iteration. In a partitioned DSM, a task is part of a sequential group if its row contains a mark just below the diagonal. Two or more tasks are parallel if there are no marks linking them. As noted, coupled tasks are identified by above-diagonal marks. Figure 12.3 shows how the DSM reveals all three types of tasks.

More sophisticated use of the DSM method has been a subject of research at MIT in the 1990s. Much of this work has applied the method to larger projects and to the development of complex systems such as automobiles and airplanes. Analytical methods have been developed to help understand the effects of complex task coupling (Smith and Eppinger, 1997a, 1997b); to predict the distribution of possible project completion times and costs (Browning and Eppinger, 1998; Carrascosa et al., 1998); and to help plan organization designs based on product architectures (Eppinger, 1997).

DSM practitioners have found that creative uses of the DSM's graphical display of project task relationships can be highly insightful for project managers in both the planning and execution phases. The chapter Appendix describes a larger DSM model in which several overlapping phases of coupled development activities are represented.

Gantt Charts

The traditional tool for representing the timing of tasks is the Gantt chart. Figure 12.4 shows a Gantt chart for the Cheetah project. The chart contains a horizontal time line created by drawing a horizontal bar representing the start and end of each task. The filled-in portion of each bar represents the fraction of the task that is complete. The vertical line in Figure 12.4 shows the current date, so we can

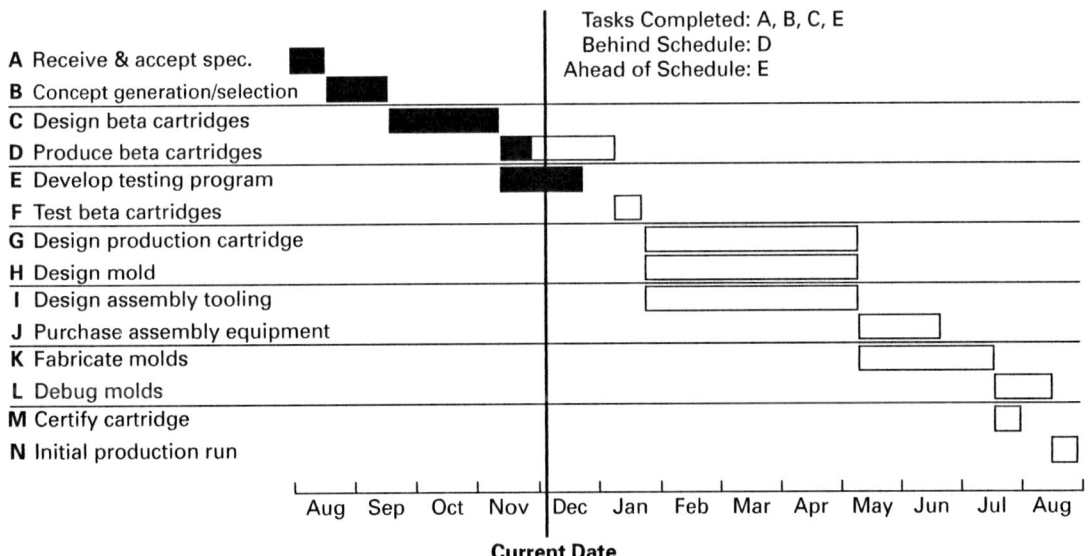

Figure 12.4 Gantt chart for the Cheetah project.

observe directly that task D is behind schedule, while task E is ahead of schedule.

A Gantt chart does not explicitly display the dependencies among tasks. Task dependencies constrain, but do not fully determine, the timing of the tasks. The dependencies dictate which tasks must be completed before others can begin (or finish, depending on the nature of the dependency) and which tasks can be completed in parallel. When two tasks overlap in time on a Gantt chart, they may be parallel, sequential, or iteratively coupled. Parallel tasks can be overlapped in time for convenience in project scheduling because they do not depend on one another. Sequential tasks might be overlapped in time, depending on the exact nature of the information dependency, as described below in the section on accelerating projects. Coupled tasks must be overlapped in time because they need to be addressed simultaneously or in an iterative fashion.

PERT Charts

PERT (program evaluation and review technique) charts explicitly represent both dependencies and timing, in effect combining some of the information contained in the DSM and Gantt chart. While there are many forms of PERT charts, we prefer the "activities on nodes" form of the chart, which corresponds to the block diagrams that most people are familiar with. The PERT chart for the Cheetah project is shown in Figure 12.5. The blocks in the PERT chart are labeled with both the task and its expected duration. Note that the PERT representation does not allow for loops or feedback and so cannot explicitly show iterative coupling. As a result, the coupled tasks G, H, and I are grouped together into one task. The graphical convention of PERT charts is that all links between tasks must proceed from left to right, indicating the temporal sequence in which tasks can be completed. When the blocks are sized to represent the duration of tasks, as in a Gantt chart, then a PERT diagram can also be used to represent a project schedule.

The Critical Path

The dependencies among the tasks in a PERT chart, some of which may be arranged sequentially and some of which may be arranged in parallel, lead to the concept of a *critical path*. The critical path is the longest chain of dependent events. This is the single sequence of tasks whose combined required times define the minimum possible completion time for the entire set of tasks. Consider for example the Cheetah project represented in Figure 12.5. Either the sequence C-D-F or the sequence C-E-F defines how much

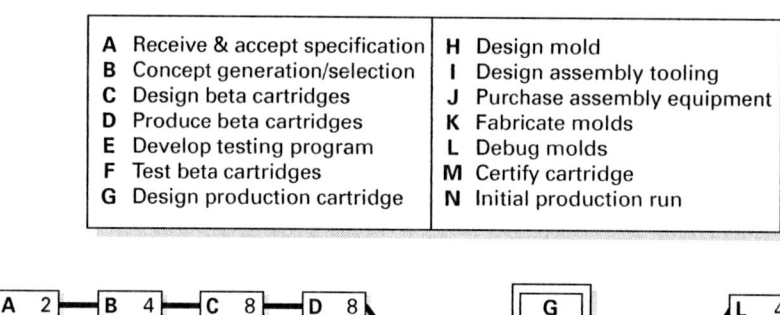

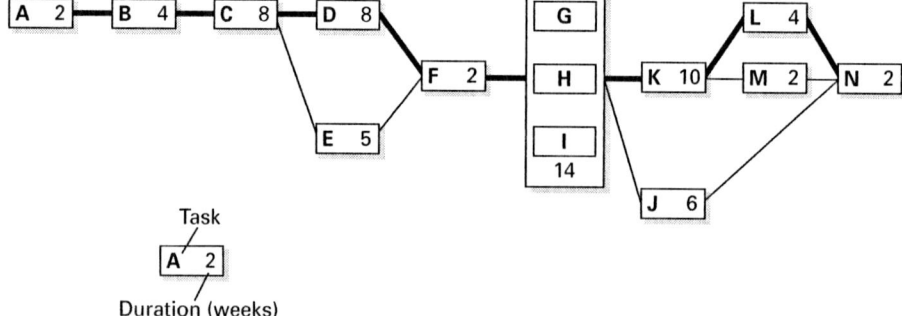

Figure 12.5 PERT chart for the Cheetah project. The critical path is designated by the thicker lines connecting tasks. Note that tasks G, H, and I are grouped together because the PERT representation does not depict coupled tasks explicitly.

time is required to complete the four tasks C, D, E, and F. In this case, the path C-D-F requires 18 weeks and the path C-E-F requires 15 weeks, so the critical path for the whole project includes C-D-F. The critical path for the project is denoted by the thick lines in Figure 12.5. Identifying the critical path is important because a delay in any of these *critical tasks* would result in an increase in project duration. All other paths contain some *slack,* meaning that a delay in one of the noncritical tasks does not automatically create a delay for the entire project. Figure 12.4 shows that task D is behind schedule. Because task D is on the critical path, this delay, if not corrected, will result in a delay of the completion of the entire project.

Several software packages are available for producing Gantt charts and PERT charts; these programs can also compute the critical path.

12.2 | BASELINE PROJECT PLANNING

The project plan is the road map for the remaining development effort. The plan is important in coordinating the remaining tasks and in estimating the required development resources and development time. Some measure of project planning occurs at the earliest stages of product development, but the importance of the plan is highest at the end of the concept development phase, just before significant development resources are committed. This section presents a method for creating a *baseline project plan.* After establishing this baseline, the team considers whether it should modify the plan to change the planned development time, budget, or project scope. The results of the concept development phase plus the project plan make up the *contract book.*

The Contract Book

We recommend that a contract book be used to document the project plan and the results of the concept development phase of the development process. The concept of a contract book is detailed by Wheelwright and Clark (1992). The word *contract* is used to emphasize that the document represents an agreement between the development team and the senior management of the company about project goals, direction, and resource requirements. The book is sometimes actually signed by the key team members and the senior managers of the company. A table of contents for a contract book is shown in Figure 12.6, along with references to the chapters in this book where some of these contents are discussed.

Below we discuss elements of the project plan: the project task list, team staffing and organization, the project schedule, the project budget, and the project risk areas.

Item	Approximate Pages	See Chapter(s)
Mission Statement	1	3
Customer Needs List	1-2	4
Competitive Analysis	1-2	3, 4, 5, 7, 8
Product Specifications	1-3	5
Sketches of Product Concept	1-2	6, 10
Concept Test Report	1-2	8
Sales Forecast	1-3	8, 13
Economic Analysis/Business Case	1-3	13
Manufacturing Plan	1-5	11
Project Plan		
Task List	1-5	14
Design Structure Matrix	2-3	14
Team Staffing and Organization	1	14
Schedule (Gantt and/or PERT)	1-2	14
Budget	1	14
Risk Areas	1	14
Project Performance Measurement Plan	1	14
Incentives	1	14
	Total 18-38 Pages	

Figure 12.6
Table of contents of a contract book for a project of moderate complexity.

Project Task List

We have already introduced the idea that a project consists of a collection of tasks. The first step in planning a project is to list the tasks which make up the project. For most product development projects the team will not be able to list every task in great detail; too much uncertainty remains in the subsequent development activities. However, the team will be able to list its best estimate of the remaining tasks at a general level of detail. To be most useful during project planning, the task list should contain from 50 to 200 items. For small projects, such as the development of a hand tool, each task may correspond, on average, to a day or two of work for a single individual. For medium-sized projects, such as the development of a computer printer, each task may correspond to a week of work for a small group of people. For a large project, such as the development of an automobile, each task may correspond to one or more months of efforts for an entire division or sub-team. For large projects, each of the tasks identified at this level may be treated as its own development project with its own project plan.

An effective way to tackle the generation of the task list is to consider the tasks in each of the remaining phases of development. For our generic development process, the phases remaining after concept development are system-level design, detail design, testing and refinement, and production ramp-up. (See Chapter 2, Development Processes and Organizations.) In some cases, the current effort will be very similar to a previous project. In these cases, the list of tasks from the previous project is a good starting point for the new task list. The Cheetah project was very similar to dozens of previous efforts. For this reason, the team had no trouble identifying the project tasks. (Its challenge was to complete them quickly.)

After listing all of the tasks, the team estimates the effort required to complete each task. Effort is usually expressed in units of person-hours, person-days, or person-weeks, depending on the size of the project. Note that these estimates reflect the "actual working time" that members of the development team would have to apply to the task and not the "elapsed calendar time" the team expects the task to require. Because the speed with which a task is completed has some influence on the total amount of effort that must be applied to the task, the estimates embody preliminary assumptions about the overall project schedule and how quickly the team will attempt to complete tasks. These estimates are typically derived from past experience or the judgment of experienced members of the development team. A task list for the Cheetah project is shown in Figure 12.7.

Task	Estimated Person-Weeks
Concept Development	
Receive and accept specification	8
Concept generation/selection	16
Detail Design	
Design beta cartridges	62
Produce beta cartridges	24
Develop testing program	24
Testing and Refinement	
Test beta cartridges	20
Design production cartridge	56
Design mold	36
Design assembly tooling	24
Purchase assembly equipment	16
Fabricate molds	16
Debug molds	24
Certify cartridge	12
Production Ramp-up	
Initial production run	16
Total	354

Figure 12.7
Task list for the Cheetah project. (This task list is abbreviated for clarity; the actual list contained over 100 tasks.)

Team Staffing and Organization

The project team is the collection of individuals who complete project tasks. Whether or not this team is effective depends on a wide variety of individual and organizational factors. Smith and Reinertsen (1991) propose seven criteria as determinants of the speed with which a team will complete product development; in our experience these criteria predict many of the other dimensions of team performance as well:

1. There are 10 or fewer members of the team.
2. Members volunteer to serve on the team.
3. Members serve on the team from the time of concept development until product launch.
4. Members are assigned to the team full-time.
5. Members report directly to the team leader.
6. The key functions, including at least marketing, design, and manufacturing, are on the team.
7. Members are located within conversational distance of each other.

While few teams are staffed and organized ideally, these criteria raise several key issues: How big should the team be? How should the team be organized relative to the larger enterprise? Which functions should be represented on the team? How can the development team of a very large project exhibit some of the agility of a small team? Here we address the issues related to team size. Chapter 1, Introduction, and Chapter 2, Development Processes and Organizations, address some of the other team and organizational issues.

The minimum number of people required on the project team can be estimated by dividing the total estimated time to complete the project tasks by the planned project duration. For example, the estimated task time for the Cheetah project was 354 person-weeks. The team hoped to complete the project in 12 months (or about 50 weeks), so the minimum possible team size would be seven people. All other things being equal, small teams seem to be more efficient than large teams, so the ideal situation would be to have a team made up of the minimum number of people, each dedicated 100 percent to the project.

Three factors make realizing this ideal difficult. First, specialized skills are often required to complete the project. For example, one of the Cheetah tasks was to design molds. Mold designers are highly specialized, and the team could not use a mold designer for a full year. Second, one or more key team members may have other unavoidable responsibilities. For example, one of the engineers on the Cheetah project was responsible for assisting in the production ramp-up of a previous project. As a result, she was only able to commit half of her time to the Cheetah project initially. Third, the work required to complete tasks on the project is not constant over time. In general, the work requirement increases steadily until the beginning of production ramp-up and then begins to taper off. As a result, the team will generally have to grow in size as the project progresses in order to complete the project as quickly as possible.

After considering the need for specialized skills, the reality of other commitments of the team members, and the need to accommodate an increase and subsequent decrease in workload, the project leader, in consultation with his or her management, identifies the full project staff and approximately when each person will join the team. When possible, team members are identified by name, although in some cases they will be identified only by area of expertise (e.g., mold designer, industrial designer). The project staffing for the Cheetah project is shown in Figure 12.8.

Project Schedule

The project schedule is the merger of the project tasks and the project timeline. The schedule identifies when major project milestones are expected to occur and when each project task is expected to begin and end. The team uses this schedule to track progress and to orchestrate the exchange of materials and information between individuals. It is therefore important that the schedule is viewed as credible by the entire project team.

We recommend the following steps to create a baseline project schedule:

1. Use the DSM or PERT chart to identify the dependencies among tasks.
2. Position the key project milestones along a timeline in a Gantt chart.
3. Schedule the tasks, considering the project staffing and other critical resources.
4. Adjust the timing of the milestones to be consistent with the time required for the tasks.

Person	Month:	1	2	3	4	5	6	7	8	9	10	11	12
Team Leader		100	100	100	100	100	100	100	100	100	100	100	100
Schedule Coordinator		25	25	25	25	25	25	25	25	25	25	25	25
Customer Liaison		50	50	50	50	25	25	25	25	25	25	25	25
Mechanical Designer 1		100	100	100	100	100	100	100	100	50	50	50	50
Mechanical Designer 2			50	100	100	100	100	100	100	50			
CAD Technician 1			50	100	100	100	100	100	100	100	50	50	50
CAD Technician 2					50	100	100	100	100	100	50		
Mold Designer 1		25	25	25	25	100	100	100	100	25	25	25	
Mold Designer 2						100	100	100	100				
Assembly Tool Designer		25	25	25	25	100	100	100	100	100	100	50	50
Manufacturing Engineer		50	50	100	100	100	100	100	100	100	100	100	100
Purchasing Engineer			50	50	100	100	100	100	100	100	100	100	100

Figure 12.8
Project staffing for the Cheetah project. Numbers shown are approximate percentages of full time.

Project milestones are useful as anchor points for the scheduling activity. Common milestones include design reviews (also called phase reviews or design gates), comprehensive prototypes (e.g., alpha prototype, beta prototype), and trade shows. Because these events typically require input from almost everyone on the development team, they serve as powerful forces for integration and act as anchor points on the schedule. Once the milestones are laid out on the schedule, the tasks can be arranged between these milestones.

The Cheetah schedule was developed by expanding the typical project phases into a set of approximately 100 tasks. The major milestones were the concept approval, the testing of beta prototype cartridges, the trade show demonstration, and production ramp-up. Relationships among these activities and the critical path were documented using a combined PERT/Gantt chart.

Project Budget

Budgets are customarily represented with a simple spreadsheet, although many companies have standard budgeting forms for requests and approvals. The major budget items are staff, materials and services, project-specific facilities, and spending on outside development resources.

For most projects the largest budget item is the cost of staff. For the Cheetah project, personnel charges made up 80 percent of the total budget. The personnel costs can be derived directly from the staffing plan by applying the *loaded* salary rates to the estimated time commitments of the staff on the project. Loaded salaries include employee benefits and overhead and are typically between two and three times the actual salary of the team member. Many companies use only one or two different rates to represent the cost of the people on a project. Average staff costs for product development projects range from $2,000 to $5,000 per person-week. For the Cheetah project, assuming an average cost of $3,000 per person-week, the total cost for the 354 person-weeks of effort would be $1,062,000.

Early in the development project, uncertainty of both timing and costs are high and the forecasts may only be accurate within 30 to 50 percent. In the later stages of the project the program uncertainty is reduced to perhaps 5 percent to 10 percent. For this reason some margin should be added to the budget as a contingency. A summary of the Cheetah project budget is shown in Figure 12.9.

Project Risk Areas

Projects rarely proceed exactly according to plan. Some of the deviations from the plan are minor and can be accom-

Item	Amount
Staff salaries	
354 person-weeks @ $3,000/week	$1,062,000
Materials and Services	125,000
Prototype Molds	75,000
Outside Resources, Consultants	25,000
Travel	50,000
Subtotal	$1,337,000
Contingency (20%)	$267,400
Total	$1,604,400

Figure 12.9 Summary budget for the Cheetah project. The production tooling and equipment are accounted for as manufacturing costs rather than as part of the development project budget. (Kodak figures are disguised and listed here only for illustration.)

modated with little or no impact on project performance. Other deviations can cause major delays, budget overruns, poor product performance, or high manufacturing costs. Often the team can assemble, in advance, a list of what might go wrong, that is, the areas of risk for the project.

We recommend that after identifying each risk, the team assess the level of the risk and then identify the actions the team will take to minimize the risk. In addition to pushing the team to work to minimize risk, the explicit identification of risk during the project planning activity helps to minimize the number of surprises the team will have to communicate to its senior management later in the project. The risk areas for the Cheetah project are shown in Figure 12.10.

Modifying the Baseline Plan

The baseline project plan embodies assumptions about how quickly the project should be completed, about the performance and cost goals for the product, and about the resources to be applied to the project. After completing a baseline plan, the team should consider whether some of these assumptions should be revisited. In particular, the team can usually choose to trade off development time, development cost, product manufacturing cost, product performance, and risk. For example, a project can sometimes be completed more quickly by spending more money. Some of these trade-offs can be explored quantitatively using the economic analysis techniques described in Chapter 13, Product Development Economics. The most common desired modification to the baseline plan is to compress the schedule. For this reason, we devote the next section to ways the team can accelerate the project.

Risk	Risk Level	Actions to Minimize Risk
Change in customer specifications.	Moderate	• Involve the customer in process of refining specifications. • Work with the customer to estimate time and cost penalties of changes.
Poor feeding characteristics of cartridge design	Low	• Build early functional prototype from machined parts. • Test prototype in microfilm machine.
Delays in mold making shop.	Moderate	• Reserve 25% of shop capacity for May-July.
Molding problems require rework of mold	High	• Involve mold maker and mold designer in the part design. • Perform mold filling computer analysis. • Establish design rules for part design. • Choose materials at end of concept development phase.

Figure 12.10 Risk areas for the Cheetah project.

12.3 | ACCELERATING PROJECTS

Product development time is often the dominant concern in project planning and execution. This section provides a set of guidelines for accelerating product development projects. Most of these guidelines are applicable at the project planning stage, although a few can be applied throughout a development project. Accelerating a project before it has begun is much easier than trying to expedite a project that is already underway.

The first set of guidelines applies to the project as a whole.

• *Start the project early.* Saving a month at the beginning of a project is just as helpful as saving a month at the end of a project, yet teams often work with little urgency before development formally begins. For example, the meeting to approve a project plan and review a contract book is often delayed for weeks because of difficulty in scheduling a meeting with senior managers. This delay at the beginning of a project costs exactly as much time as the same delay during production ramp-up. The easiest way to complete a project sooner is to start it early.

• *Manage the project scope.* There is a natural tendency to add additional features and capabilities to the product as development progresses. Some companies call this phenomenon "feature creep" or "creeping elegance," and in time-sensitive contexts it may result in an elegant product without a market. Disciplined teams and organizations are able to "freeze the design" and leave incremental improvements for the next generation of the product.

• *Facilitate the exchange of essential information.* As shown in the DSM representation, a tremendous amount of information must be transferred within the product development team. Every task has one or more internal customers for the information it produces. For small teams, frequent exchange of information is quite natural and is facilitated by team meetings and colocation of team members. Larger teams may require more structure to promote rapid and frequent information exchange. Blocks of coupled tasks revealed by the DSM identify the specific needs for intensive information exchange. Computer networks and emerging software tools offer some promise for enhancing this exchange within larger development teams.

The second set of guidelines is aimed at decreasing the time required to complete the tasks on the critical path. These guidelines arise from the fact that the only way to reduce the time required to complete a project is to shorten the critical path. Note that a decision to allocate additional resources to shortening the critical path should be based on the value of accelerating the entire project. For some projects, time reductions on the critical path can be worth hundreds of thousands, or even millions, of dollars per week.

• *Complete individual tasks on the critical path more quickly.* The benefit of recognizing the critical path is that the team can focus its efforts on this vital sequence of tasks. The critical path generally represents only a small fraction of the total project effort, and so additional spending on completing a critical task more quickly can usually quite easily be justified. Sometimes completing critical tasks more quickly can be achieved simply by identifying a task as critical so that it gets special attention, starts earlier, and is not interrupted. Note that the accelerated completion of a critical task may cause the critical path to shift to include previously noncritical tasks.

- *Aggregate safety times.* The estimated duration of each task in the project generally includes some amount of "safety time." This time accounts for the many normal but unpredictable delays which occur during the execution of each task. Common delays include: waiting for information and approvals, interruptions from other tasks or projects, and tasks being more difficult than anticipated. Goldratt (1997) estimates that built-in safety doubles the nominal duration of tasks. Although safety time is added to the expected task duration to account for random delays, these estimates become targets during execution of the tasks, which means that tasks are rarely completed early and many tasks overrun. Goldratt recommends removing the safety time from each task along the critical path and aggregating all of the safety time from the critical path into a single *project buffer* placed at the end of the project schedule. Because the need to extend task duration occurs somewhat randomly, only some of the tasks will actually need to utilize time from the project buffer. Therefore, a single project buffer can be smaller than the sum of the safety times that would be included in each estimate of task duration, and the critical path may be completed sooner. In practice, the project buffer may only need to start with time equal to half of the shortened critical path duration. Goldratt has developed these ideas into a project management method called *Critical Chain.* In addition to the project buffer, the method uses *feeder buffers* to protect the critical path from delays where noncritical tasks feed into the critical path. Each feeder buffer aggregates the safety times of the tasks on a noncritical path. Figure 12.11 illustrates the use of project and feeder buffers.

- *Eliminate some critical path tasks entirely.* Scrutinize each and every task on the critical path and ask whether it can be removed or accomplished in another way.
- *Eliminate waiting delays for critical path resources.* Tasks on the critical path are sometimes delayed by waiting for a busy resource. The waiting time is frequently longer than the actual time required to complete the task. Delays due to waiting are particularly prominent when procuring special components from suppliers. Sometimes such delays can be avoided by ordering an assortment of materials and components in order to be sure to have the right items on hand, or by purchasing a fraction of the capacity of a vendor's production system in order to expedite the fabrication of prototype parts. These expenses may make perfect economic sense in the context of the overall development project, even though the expenditure may seem extravagant when viewed in isolation. In other cases, administrative tasks such as purchase order approvals may become bottlenecks. Because in past cartridge development projects periodic budget approvals had caused delays, the Cheetah project leader began early to pursue aggressively the necessary signatures so as not to hold up the activities of the entire team.
- *Overlap selected critical tasks.* By scrutinizing the relationships between sequentially dependent tasks on the critical path, the tasks can sometimes be overlapped or executed in parallel. In some cases, this may require a significant redefinition of the tasks or even changes to the architecture of the product. (See Chapter 9, Product Architecture, for more details on dependencies arising from the architecture of the product.) In other cases, overlapping entails simply transferring partial information earlier and/or more frequently between nominally sequential tasks or freezing the critical upstream information earlier. Krishnan (1996) provides a framework for choosing various overlapping strategies.

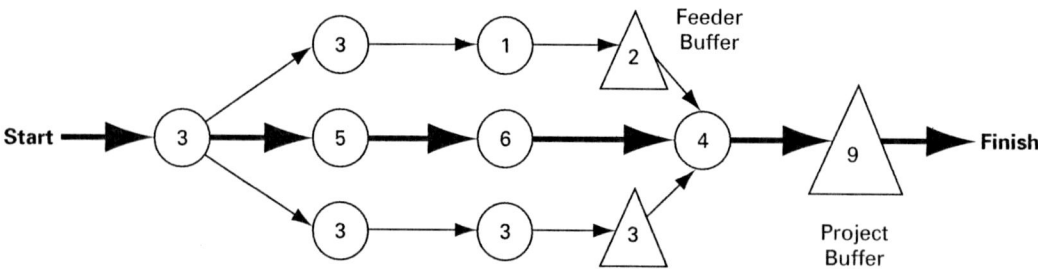

Figure 12.11 The Critical Chain method aggregates the safety time along the critical path into the project buffer. Feeder buffers protect the critical path from delays. In this illustration, nominal task durations (in days) are given for each task, and the critical path is shown with thicker arrows linking the critical tasks.

- *Pipeline large tasks.* The strategy of *pipelining* is applied by breaking up a single large task into smaller tasks whose results can be passed along as soon as they are completed. For example, the process of finding and qualifying the many vendors which supply the components of a product can be time-consuming and can even delay the production ramp-up if not completed early enough. Instead of waiting until the entire bill of materials is complete before the purchasing department begins qualifying vendors, purchasing could qualify vendors as soon as each component is identified. Pipelining in effect allows nominally sequential tasks to be overlapped.
- *Outsource some tasks.* Project resource constraints are common. When a project is constrained by available resources, assigning tasks to an outside firm or to another group within the company may prove effective in accelerating the overall project.

The final set of guidelines is aimed at completing coupled tasks more quickly. Recall that coupled tasks are those which must be completed simultaneously or iteratively because they are mutually dependent.

- *Perform more iterations quickly.* Much of the delay in completing coupled tasks is in passing information from one person to another and in waiting for a response. If the iteration cycles can be completed at a higher frequency, then the coupled tasks can sometimes be competed more quickly. Faster iterations can be achieved through faster and more frequent information exchanges. In the Cheetah project, the mechanical engineer would work closely with the mold designer, who would in turn work closely with the mold maker. In many cases, these three would share a single computer terminal for the purpose of exchanging ideas about how the design was evolving from their three different perspectives.
- *Decouple tasks to avoid iterations.* Iterations can often be reduced or eliminated by taking actions to decouple tasks. For example, by clearly defining an interface between two interacting components early in the design process, the subsequent design of the two components can proceed independently and in parallel. The definition of the interface may take some time in advance, but the avoidance of time-consuming iterations may result in net time savings. (See Chapter 9, Product Architecture, for a discussion of establishing interfaces in order to allow the independent development of components.)
- *Consider sets of solutions.* Iterations involve the exchange of information about the evolving product design. Rather than exchanging point-value estimates of

design parameters, in some cases the use of ranges or sets of values may facilitate faster convergence of coupled tasks. Researchers have recently described the application of such set-based approaches to concurrent engineering at Toyota (Sobek et al., 1999).

12.4 | PROJECT EXECUTION

Smooth execution of even a well-planned project requires careful attention. Three problems of project execution are particularly important: (1) What mechanisms can be used to coordinate tasks? (2) How can project status be assessed? and (3) What actions can the team take to correct for undesirable deviations from the project plan? We devote this section to these issues.

Coordination Mechanisms

Coordination among the activities of the different members of the team is required throughout a product development project. The need for coordination is a natural outgrowth of dependencies among tasks. Coordination needs also arise from the inevitable changes in the project plan caused by unanticipated events and new information. Difficulties in coordination can arise from inadequate exchanges of information and from organizational barriers to cross-functional cooperation. Here are several mechanisms used by teams to address these difficulties and facilitate coordination.

- *Informal communication:* A team member engaged in a product development project may communicate with other team members dozens of times per day. Many of these communications are informal; they involve a spontaneous stop by someone's desk or a telephone call to solicit a piece of information. Good informal communication is one of the mechanisms most useful in breaking down individual and organizational barriers to cross-functional cooperation. Informal communication is dramatically enhanced by locating the core members of the development team in the same work space. Allen (1977) has shown that communication frequency is inversely related to physical separation and falls off rapidly when people are located more than a few meters from one another (Figure 12.12). In our experience, electronic mail and, to a lesser extent, voice mail also provide effective means of fostering informal communication among people who are already well acquainted with one another.

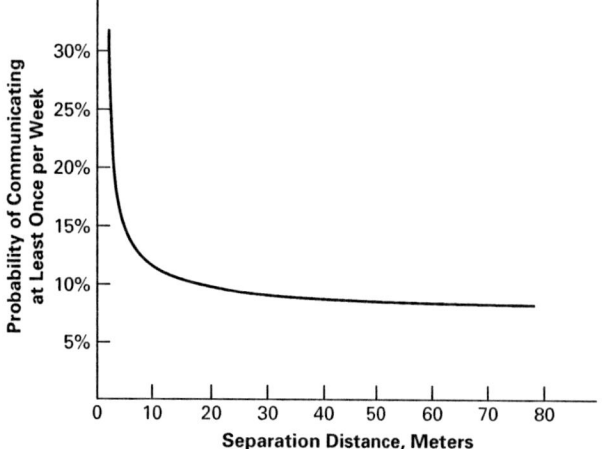

Figure 12.12 Communication frequency versus separation distance. This relationship shown is for individuals with an organizational bond, such as belonging to the same product development team.

- *Meetings:* The primary formal communication mechanism for project teams is meetings. Most teams meet formally at least once each week. Many teams meet twice each week, and some teams meet every day. Teams located in the same work space need fewer formal meetings than those whose members are geographically separated. Time spent exchanging information in meetings is time not spent completing other project tasks. In order to minimize the amount of time wasted in meetings, some teams that hold frequent meetings meet standing up to emphasize that the meeting is intended to be quick. Other techniques for controlling the length of meetings include preparing a written agenda, appointing someone to run the meeting, and holding the meeting just before lunchtime or near the end of the day when people are anxious to leave. We recommend that team meetings be held at a regular time and place so that no extra effort is expended in scheduling the meeting and in informing the team of its time and location.
- *Schedule display:* The most important information system in project execution is the project schedule, usually in the form of a PERT or Gantt chart. Most successful projects have a single person who is responsible for monitoring the schedule. On small projects, this is usually the team leader. Larger projects generally have a designated person other than the project leader who watches and updates the schedule regularly. On the Cheetah project, Kodak provided a part-time project analyst who kept the schedule current on a weekly

basis and reported to the project team leader. The team members understood the importance of accurate schedule projections and were very cooperative in this effort. Schedule updates are usually displayed in Gantt chart form (Figure 12.4).

- *Weekly updates:* The *weekly status memo* is written by the project leader and is distributed on paper, in electronic form, or even by voice mail to the entire extended project team, usually on Friday or over the weekend. The memo is usually one or two pages long and lists the key accomplishments, decisions, and events of the past week. It also lists the key events of the coming week. It is sometimes accompanied by an updated schedule.
- *Incentives:* Some of the most basic organizational forms, such as functional organizations which use functional performance reviews, may inhibit the productive collaboration of team members across functions. The implementation of project-based performance measures creates incentives for team members to contribute more fully to the project. Having both a project manager and a functional manager contribute to individual performance reviews leading to promotions, merit increases, and bonuses, sends a strong message that project results are highly valued. (See Chapter 2, Development Processes and Organizations, for a discussion of various organizational forms, including project, functional, and matrix organizations.)
- *Process documents:* Each of the methods presented in this book also has an associated information system which assists the project team in making decisions and provides documentation. (By information systems we mean all of the structured means the team uses to exchange information, not only the computer systems used by the team.) For example, the concept selection method uses two concept selection matrices to both document and facilitate the selection process. Similarly, each of the other information systems serves both to facilitate the logical execution of the process step and to document its results. Figure 12.13 lists some of the important information systems used at the various stages of the development process.

Assessing Project Status

Project leaders and senior managers need to be able to assess project status to know whether corrective actions are warranted. In projects of modest size (say, fewer than 50 people) project leaders are fairly easily able to assess the status of the project. The project leader assesses project sta-

Development Activity	Information Systems Used
Product planning	Product segment map
	Technology roadmap
	Product-process change matrix
	Aggregate resource plan
	Product plan
	Mission statement
Customer needs identification	Customer needs lists
Concept generation	Function diagrams
	Concept classification tree
	Concept combination table
	Concept descriptions and sketches
Concept selection	Concept screening matrix
	Concept scoring matrix
Product specifications	Needs-metrics matrix
	Competitive benchmarking charts
	Specifications lists
System-level design	Schematic diagram
	Geometric layout
	Differentiation plan
	Commonality plan
Detailed design	Bill of materials
	Prototyping plan
Industrial design	Aesthetic/ergonomic importance survey
Product development economics	NPV analysis spreadsheet
Project management	Contract book
	Task list
	Design structure matrix
	Gantt chart
	PERT chart
	Staffing matrix
	Risk analysis
	Weekly status memo
	Buffer report
	Postmortem project report

Figure 12.13 Information systems which facilitate product development decision making, team consensus, and the exchange of information.
(Courtesy of FIAT Auto)

tus during formal team meetings, by reviewing the project schedule, and by gathering information in informal ways. The leader constantly interacts with the project team, meets regularly with individuals to work through difficult problems, and is able to observe all of the information systems of the project. A team may also engage an expert from outside the core team to review the status of the project. The

goal of these reviews is to highlight areas of risk and to generate ideas for addressing these risk areas.

Project reviews, conducted by senior managers, are another common method of assessing progress. These reviews tend to correspond to the end of each phase of development and are key project milestones. These events serve not only to inform senior managers of the status of a project but also to bring closure to a wide variety of development tasks. While these reviews can be useful milestones and can enhance project performance, they can also hinder performance. Detrimental results arise from devoting too much time to preparing formal presentations, from delays in scheduling reviews with busy managers, and from excessive meddling in the details of the project by those reviewing the project.

The Critical Chain method uses a novel approach to monitoring the project schedule. By simply monitoring the project buffer and the feeder buffers of the project (described briefly above), the project manager can quickly assess the criticality of each path and the estimated project completion time. If tasks consume the project buffer faster than the critical path is being completed, the project runs the risk of slipping the end date. A buffer report therefore provides a concise update on the project status in terms of progress of the critical path and its feeder paths.

Corrective Actions

After discovering an undesirable deviation from the project plan, the team attempts to take corrective action. Problems almost always manifest themselves as potential schedule slippage, and so most of these corrective actions relate to arresting potential delays. Some of the possible actions include:

- *Changing the timing or frequency of meetings:* Sometimes a simple change from weekly to daily meetings increases the "driving frequency" of the information flow among team members and enables more rapid completion of tasks. This is particularly true of teams that are not already colocated (although if the team is highly dispersed geographically, meetings can consume a great deal of travel time). Sometimes simply moving a weekly meeting from a Tuesday morning to a Friday afternoon increases the urgency felt by the team to "get it done this week."

- *Changing the project staff:* The skills, capabilities, and commitment of the members of the project team in large measure determine project performance. When the project team is grossly understaffed, performance can sometimes be increased by adding the necessary staff. When the project team is overstaffed, performance can sometimes be increased by removing staff. Note that adding staff in a panic at the end of a project can lead to delays in project completion because the increased coordination requirements may outweigh the increase in human resources.

- *Locating the team together physically:* If the team is geographically dispersed, one way to increase project performance is to locate the team in the same work space. This action invariably increases communication among the team members. Some measure of "virtual colocation" is possible with electronic mail, video conferencing, and other network-based collaboration tools.
- *Soliciting more time and effort from the team:* If some team members are distributing their efforts among several projects, project performance may be increased by relieving them of other responsibilities. Needless to say, high-performance project teams include team members who regularly deliver more than a 40-hour work week to the project. If a few critical tasks demand extraordinary effort, most committed teams are willing to devote a few weeks of 14-hour days to get the job done. However, 60- or 70-hour weeks cannot reasonably be expected from most team members for more than a few weeks without causing fatigue and burnout.
- *Focusing more effort on the critical tasks:* By definition, only one sequence of tasks forms the critical path. When the path can be usefully attacked by additional people, the team may choose to temporarily drop some or all other noncritical tasks in order to ensure timely completion of the critical tasks.
- *Engaging outside resources:* The team may be able to retain an outside resource such as a consulting firm or a supplier to perform some of the development tasks. Outside firms are typically fast and relatively economical when a set of tasks can be clearly defined and when coordination requirements are not severe.
- *Changing the project scope or schedule:* If all other efforts fail to correct undesirable deviations from the project plan, then the team must either narrow the scope of the project or extend (slip) the project schedule. These changes are necessary to maintain a credible and useful project plan.

12.5 | POSTMORTEM PROJECT EVALUATION

An evaluation of the project's performance after it has been completed is useful for both personal and organizational improvement. This review is often called a *post-*

mortem project evaluation, although more friendly names are appropriate (Smith, 1996). The postmortem evaluation is usually an open-ended discussion of the strengths and weaknesses of the project plan and execution. This discussion is sometimes facilitated by an outside consultant or by someone within the company who was not involved in the project. Several questions help to guide the discussion:

- Did the team achieve the mission articulated in the mission statement?
- Which aspects of project performance (development time, development cost, product quality, manufacturing cost) were most positive?
- Which aspects of project performance were most negative?
- Which tools, methods, and practices contributed to the positive aspects of performance?
- Which tools, methods, and practices detracted from project success?
- What problems did the team encounter?
- What specific actions can the organization take to improve project performance?
- What specific technical lessons were learned? How can they be shared with the rest of the organization?

A postmortem report is then prepared as part of the formal closing of the project. These reports are used in the project planning stage of future projects to help team members know what to expect and to help identify what pitfalls to avoid. The reports are also a valuable source of historical data for studies of the firm's product development practices. Together with the project documentation, and particularly the contract book, they provide "before and after" views of each project.

For the Cheetah project, the postmortem discussion involved six members of the core team and lasted two hours. The discussion was facilitated by a consultant. The project was completed on time, despite the aggressive schedule, so much of the discussion focused on what the team had done to contribute to project success. The team agreed that the most important contributors to project success were:

- Empowerment of a team leader.
- Effective team problem solving.
- Emphasis on adherence to schedule.
- Effective communication links.
- Full participation from multiple functions.
- Building on prior experience in cartridge development.
- Use of computer-aided design (CAD) tools for communication and analysis.
- Early understanding of manufacturing capabilities.

The Cheetah team also identified a few opportunities for improvement:

- Use of three-dimensional CAD tools and plastic molding analysis tools.
- Earlier participation by the customer in the design decisions.
- Improved integration of tooling design and production system design.

12.6 | SUMMARY

Successful product development requires effective project management. Some of the key ideas in this chapter are:

- Projects consist of tasks linked to each other by dependencies. Tasks can be sequential, parallel, or coupled.
- The longest chain of dependent tasks defines the critical path, which dictates the minimum possible completion time of the project.
- The design structure matrix (DSM) can be used to represent dependencies. Gantt charts are used to represent the timing of tasks. PERT charts represent both dependencies and timing and are frequently used to compute the critical path.
- Project planning results in a task list, a project schedule, staffing requirements, a project budget, and a risk assessment. These items are key elements of the contract book.
- Most opportunities for accelerating projects arise during the project planning phase. There are many ways to complete development projects more quickly.
- Project execution involves coordination, assessment of progress, and taking action to address deviations from the plan.
- Evaluating the performance of a project encourages and facilitates personal and organizational improvement.

12.6 | REFERENCES AND BIBLIOGRAPHY

Many current resources are available on the Internet via

www.ulrich-eppinger.net

There are many basic texts on project management, although most do not focus on product development projects. PERT, CPM, and Gantt techniques are described in most project management books, including Kerzner's classic text. Kerzner also discusses project staffing, planning, budgeting, risk management, and control.

Kerzner, H., *Project Management: A Systems Approach to Planning, Scheduling, and Controlling,* sixth edition, Wiley, New York, 1998.

Several authors have written specifically about the management of product development. Wheelwright and Clark discuss team leadership and other project management issues in depth.

Wheelwright, Stephen C., and Kim B. Clark, *Revolutionizing Product Development: Quantum Leaps in Speed, Efficiency, and Quality,* The Free Press, New York, 1992.

The design structure matrix (DSM) was originally developed by Steward in the 1970s. More recently, this method has been applied to industrial project planning and assessment by Eppinger and his research group at MIT.

Steward, Donald V., *Systems Analysis and Management: Structure, Strategy, and Design,* Petrocelli Books, New York, 1981.

Eppinger, Steven D., et al., "A Model-Based Method for Organizing Tasks in Product Development," *Research in Engineering Design,* Vol. 6, No. 1, 1994, pp.1-13.

Smith, Robert P., and Steven D. Eppinger, "Identifying Controlling Features of Engineering Design Iteration," *Management Science,* Vol. 43, No. 3, March 1997, pp.276-293.

Smith, Robert P., and Steven D. Eppinger, "A Predictive Model of Sequential Iteration in Engineering Design," *Management Science,* Vol. 43, No. 8, August 1997, pp.1104-1120.

Carrascosa, Maria, Steven D. Eppinger, and Daniel E. Whitney, "Using the Design Structure Matrix to Estimate Time to Market in a Product Development Process," *ASME Design Automation Conference,* Atlanta, GA, September 1998, No. DETC98-6013.

Browning, Tyson R., and Steven D. Eppinger, "A Model for Development Project Cost and Schedule Planning," MIT Sloan School Working Paper 4050, November 1998.

Eppinger, Steven D., "A Planning Method for Integration of Large-Scale Engineering Systems," *International Conference on Engineering Design,* Tampere, Finland, August 1997, pp.199-204.

Krishnan provides a framework for overlapping nominally sequential tasks, explaining under what conditions it is better to transfer preliminary information from upstream to downstream and when it may be better to freeze the upstream task early.

Krishnan, Viswanathan, "Managing the Simultaneous Execution of Coupled Phases in Concurrent Product Development," *IEEE Transactions on Engineering Management.* Vol. 43, No. 2, May 1996, pp.210-217.

Goldratt developed the Critical Chain method of project management. This approach aggregates safety times from each task into project and feeder buffers, allowing the project to be tracked by monitoring these buffers.

Goldratt, Eliyahu M., *Critical Chain,* North River Press, Great Barrington, MA, 1997.

Smith and Reinertsen provide many ideas for accelerating product development projects, along with interesting insights on team staffing and organization.

Smith, Preston G., and Donald G. Reinertsen, *Developing Products in Half the Time,* Van Nostrand Reinhold, New York, 1991.

Sobek, Ward, and Liker present the principles of set-based concurrent engineering, in which product development teams reason about sets of possible design solutions rather than using only point-based values to describe the evolving design.

Sobek II, Durward K., Allen C. Ward, and Jeffrey K. Liker, "Toyota's Principles of Set-Based Concurrent Engineering," *Sloan Management Review.* Vol. 40, No. 2, Winter 1999, pp.67-83.

Allen has extensively studied communication in R&D organizations. This text includes the results of his seminal empirical studies of the influence of physical layout on communication.

Allen, Thomas, J., *Managing the Flow of Technology: Technology Transfer and the Dissemination of Technological Information within the R&D Organization,* MIT Press, Cambridge, MA, 1977.

Kostner offers guidance for leaders of geographically dispersed teams.

Kostner, Jaclyn, *Virtual Leadership: Secrets from the Round Table for the Multi-Site Manager,* Warner Books, New York, 1994.

Markus explains that electronic mail can facilitate rich interactions between project team members, in addition to traditional rich media such as face-to-face meetings.

Markus, M. Lynne, "Electronic Mail as the Medium of Managerial Choice," *Organization Science,* Vol. 5, No. 4, November 1994, pp502-527.

Hall presents a structured process for risk identification, analysis, and management, with application examples in software and systems engineering. (See also Kerzner, 1998.)

Hall, Elaine M., *Methods for Software Systems Development,* Addison Wesley, Reading, MA, 1998.

Smith presents a 12-step process for project review and evaluation, leading to ongoing improvement of the product development process.

Smith, Preston G., "Your Product Development Process Demands Ongoing Improvement," *Research–Technology Management,* Vol. 39, No. 2, March-April 1996, pp.37-44.

Exercises

1. The tasks for preparing a dinner (along with the normal completion times) might include:
   - Wash and cut vegetables for the salad (15 minutes).
   - Toss the salad (2 minutes).
   - Set the table (8 minutes).
   - Start the rice cooking (2 minutes).
   - Cook rice (25 minutes).
   - Place the rice in a serving dish (1 minute).
   - Mix casserole ingredients (10 minutes).
   - Bake the casserole (25 minutes).

   Prepare a DSM for these tasks.

2. Prepare a PERT chart for the tasks in Exercise 1. How fast can one person prepare this dinner? What if there were two people?

3. What strategies could you employ to prepare dinner more quickly? If you thought about dinner 24 hours in advance, are there any steps you could take to reduce the time between arriving home the next day and serving dinner?

4. Interview a project manager (not necessarily from product development). Ask him or her to describe the major obstacles to project success.

Thought Questions

1. When a task on the critical path (e.g., the fabrication of a mold) is delayed, the completion of the entire project is delayed, even though the total amount of work required to complete the project may remain the same. How would you expect such a delay to impact the total cost of the project?

2. This chapter has focused on the "hard" issues in project management related to tasks, dependencies, and schedules. What are some of the "soft," or behavioral, issues related to project management?

3. What would you expect to be some of the characteristics of individuals who successfully lead project teams?

4. Under what conditions might efforts to accelerate a product development project also lead to increased product quality and/or decreased product manufacturing costs? Under what conditions might these attributes of the product deteriorate when the project is accelerated?

Appendix: Design Structure Matrix Example

One of the most useful applications of the design structure matrix (DSM) method is to represent well-established, but complex, engineering design processes. This rich process modeling approach facilitates:

- Understanding of the existing development process.
- Communication of the process to the people involved.
- Process improvement.
- Visualization of progress during the project.

Figure 12.A1 shows a DSM model of a critical portion of the development process at a major automobile manufacturer. The model includes 50 tasks involved in the digital-mock-up (DMU) process for the layout of all of the many components in the engine compartment of the vehicle. The process takes place in six phases, depicted by the blocks of activities along the diagonal. The first two of these phases (project planning and CAD data collection) occur in parallel, followed by the development of the digital assembly model (DMU preparation). Each of the last three phases involves successively more accurate analytical verification that components represented by the digital assembly model actually fit properly within the engine compartment area of the vehicle.

In contrast to the simpler DSM model shown in Figure 12.3, where the squares on the diagonal identify sets of coupled activities, the DSM in Figure 12.A1 uses such blocks to show which activities are executed together (in parallel, sequentially, and/or iteratively) within each phase. Arrows and dashed lines represent the major iterations between sets of activities within each phase.

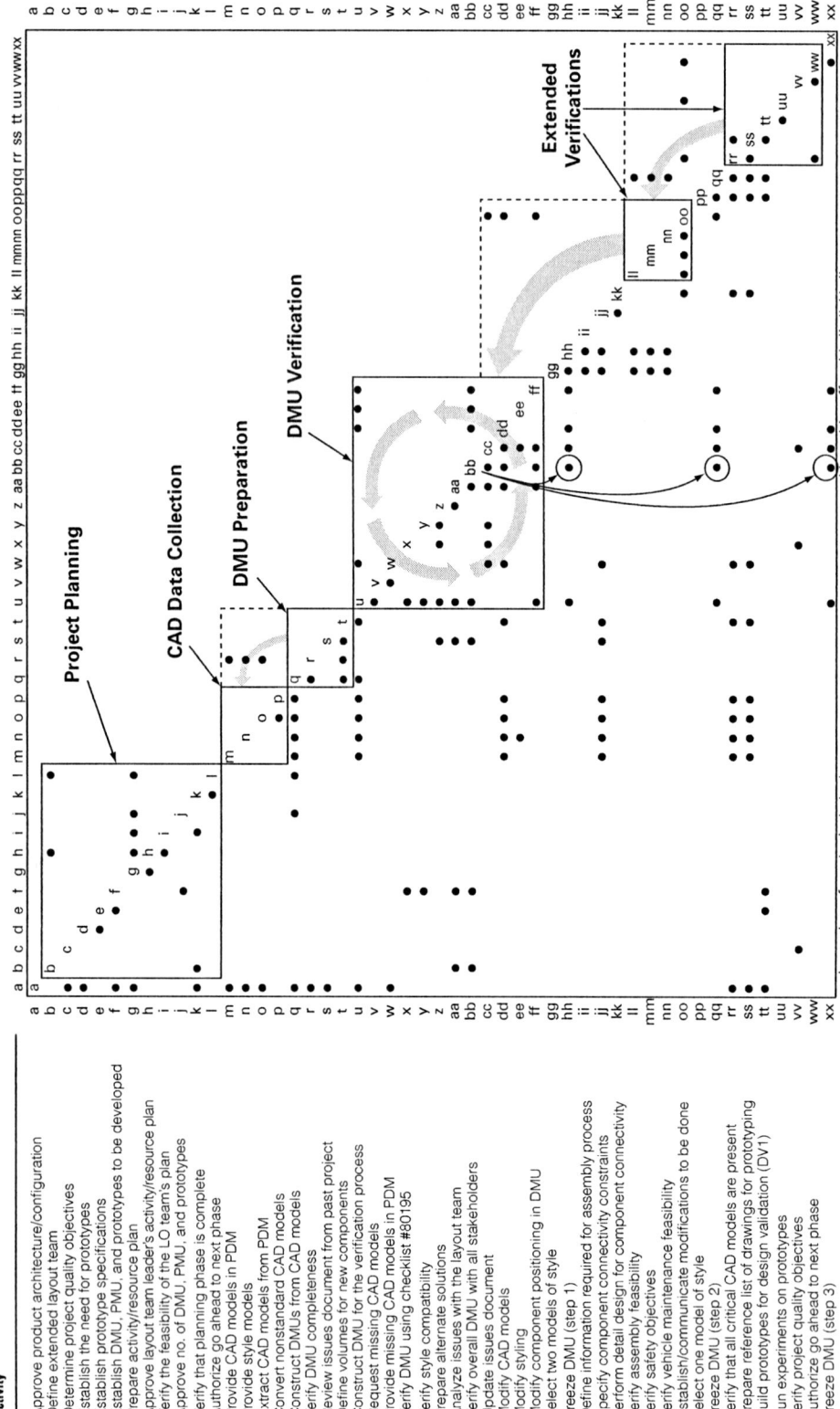

Figure 12.A1 Design structure matrix model of the digital-mock-up (DMU) process used to validate layout of the automobile's engine compartment. (Courtesy of FIAT Auto)

Index